The God/Man/World Triangle

Also by Robert Crawford

CAN WE EVER KILL? (*second enlarged edition*)

JOURNEY INTO APARTHEID

MAKING SENSE OF THE STUDY OF RELIGION

A PORTRAIT OF THE ULSTER PROTESTANTS

THE SAGA OF GOD INCARNATE (*second enlarged edition*)

WHAT IS RELIGION? (*forthcoming*)

The God/Man/World Triangle

A Dialogue between Science and Religion

Robert Crawford

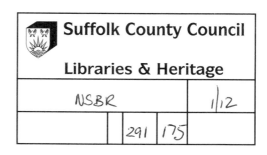

First published in Great Britain 2000 by
MACMILLAN PRESS LTD
Houndmills, Basingstoke, Hampshire RG21 6XS and London
Companies and representatives throughout the world

A catalogue record for this book is available from the British Library.

ISBN 0–333–68935–6 hardcover
ISBN 0–333–80400–7 paperback

First published in the United States of America 2000 by
ST. MARTIN'S PRESS, LLC,
Scholarly and Reference Division,
175 Fifth Avenue, New York, N.Y. 10010

ISBN 0–312–17530–2 (cloth)
ISBN 0–312–23238–1 (paperback)

The Library of Congress has cataloged the hardcover edition as follows:
Crawford, Robert G. (Robert George), 1927–
    The God/man/world triangle : a dialogue between science and religion /
Robert Crawford.
      p. cm.
    Includes bibliographical references and index.
    ISBN 0–312–17530–2
    1. Religion and science. I. Title.

BL240 .2.C73 1997
215—dc21
                                                                97–5893
                                                                CIP

First edition 1997
First paperback edition (with minor corrections) 2000

This book is printed on paper suitable for recycling and made from fully managed and sustained
forest sources.

10   9   8   7   6   5   4   3   2   1
09 08 07 06 05 04 03 02 01 00

Printed and bound in Great Britain by
Antony Rowe Ltd, Chippenham, Wiltshire

To my son Paul
who graduated in science

Contents

Introduction

This book is concerned with questions which arise when we think about mankind, the world and God. Science tells us a great deal about how the laws of the universe operate but can it tell us what it is like in itself? And, why does it exist? Is it eternal or will it cease to be? Has it a creator or can it be the cause of its own existence? If there is a God where is he and what is he like? And how are we related to him? Is it possible to experience him? What of ourselves: are we simply 'survival machines' powered and programmed by genes or some superior kind of animals not really distinct from the sub-human? Is there evidence that shows that we are distinctive? We will seek answers to these and other questions from both religion and science.

Some people contend that religion and science should be kept apart but science has impinged upon religious beliefs and religion needs to engage in accommodation otherwise it retreats into a ghetto of private feelings. We will argue that science and religion complement one another in understanding God, mankind and the world. We will oppose the view that religion is in conflict with science which dispels religious superstitions and mysteries and gives an ultimate explanation of everything. We acknowledge the prestige of science and its benefits but will argue that its explanations, while correct, are incomplete.

Science answers the question as to how things come into existence while religion tries to deal with why they exist, that is, meaning, purpose and value. But the two questions interrelate because issues arise which involve both. Perhaps that is what Einstein meant when he said, 'science without religion is lame, religion without science is blind'. There are disagreements about our triangle both within science and religion and between the two disciplines but it is possible to show that there are some agreements. Religion cannot be based on science but its views should be consonant with the scientific understanding of the world and man otherwise it fails to communicate with its society. Conversely, if science seeks to proceed without any acknowledgement of the values and purpose which religion imparts to life, its methods may become inhumane and its goals meaningless.

Another aim is to demonstrate that certain scientists in making the move from science to a materialistic perspective, that is, declaring that the only things which exist in a purposeless world are material things, are being metaphysical. The conflict is not then with science but with life viewed as materialistic rather than spiritual. Religion is not in competition with science but with a scientism that asserts it can explain everything on a mechanistic basis.

What we also aim to explain is that God, mankind and the world are interrelated and that there is both continuity and discontinuity between them. In so doing we would stress that this is not another discussion of the relationship between science and religion but an attempt to see if we can gain insights from both in our understanding of the triangle. We will be concerned mainly with physics and biology but in Chapter 6 we seek additional help from psychology. Science has developed in countries other than the Western world but it has been in the context of Christianity that the greatest achievements have been made. Usually books concentrate on this but it is essential to remember that Islam, for example, has made a contribution. Today in any dialogue it is necessary to pay attention to faiths other than Christianity not only because we live in 'a global village' but also to understand the variety of religious expression and the insights that other faiths may provide.

In the first chapter we sketch briefly how the scientific understanding of the world developed and contrast it with the religious and philosophical view of the ancient and medieval world. The old view saw God active in the world, having a purpose for it, and giving meaning and value to mankind. But the new view concentrated on the laws which described nature's behaviour. The result was a picture of the world as a machine with God as a remote spectator who only occasionally intervened.

Chapter 2 discusses the theory of evolution, the change in viewing the world as an organism, and the challenge to the design argument for the existence of God and his presence in the world. Some clergy rejected the theory but others understood it as God's method of creation. The impact of

Darwinian science and historical criticism of the scriptures caused a retreat of some theologians into the area of personal ethics and feeling, leaving the world to the study of science. Chapter 3 considers neo-Darwinism and the claim that religious faith is blind trust. We argue that this is incorrect, shows a misunderstanding of religion, and rests on a mechanistic view of mankind. The method of evolution is debated among biologists but the theory is not in doubt. We see that mystery still surrounds the origin of life.

Chapter 4 sketches briefly the Einstein revolution in physics and the change from the Newtonian view of the world. It is seen that we now live in a dynamic and interconnected universe where instead of separate enduring things, externally related to each other, we have a unified flux of interacting events. Relationship becomes very important. Taking their cue from Einstein's view of spacetime, some physicists and theologians suggest that God might be in another dimension. But we believe that God's transcendence need not be understood in spatial ways but in terms of his greatness and holiness.

In Chapter 5 we discover that science instead of decreasing mystery has increased it. Here is the strange world of the quantum with its consequences for the beginning of the universe, black holes, imaginary time, uncertainty regarding the behaviour of the smallest particles, complementarity and the role of the observer. The world appears to be more open, less deterministic, novel and dynamic which leaves room for divine initiative. We look at some of the attempts to relate this world to the presence of God within it. Since a mechanistic view of humanity has been stressed, particularly in biology, we think it necessary, in Chapter 6, to argue for the distinctiveness of mankind. Continuity with the animals is not denied but on the basis of insights drawn mainly from psychology, we believe that our mental ability, freedom of choice, values, consciousness of the self, personhood and the desire to worship, set us apart. In brief, that we are more than survival machines.

In Chapters 7 and 8 we consider what the Indian and Semitic traditions teach about the triangle. There are differences of opinion but most are saying that there is a spiritual presence in the world and we consider various models of what they think this presence may be like. Man is imperfect, deluded by material values, and in need of liberation or salvation. Chapter 9 continues the discussion of models of God – cosmic self, agent, self-emptying, human, incarnate, dual aspect, trinity, eschatological – as a prelude to finding a model in the final chapter. But are the religions simply based on feeling or are there rational arguments for the existence of God? The question is explored in Chapter 10 and the conclusion reached that while there is no proof, there is cumulative evidence for the probability of his existence. The final chapter brings together God, mankind and the world, and reaffirms our belief that there is continuity yet discontinuity between them. The most mysterious concept is God and any model we put forward can only be suggestive. Our model is that of the cosmic scientist with the world as a vast experiment. It reflects what we have learned about the world but complemented by the more personal model which arises from the religious traditions.

Because of the limitation of space imposed by a book of this size and the vastness of the areas considered, these matters could only be dealt with in a limited way but it is hoped that the attempt will stimulate discussion and lead to more extensive treatment. God does not have a gender but it is impossible to write about the concept without using gender-specific terminology. I have used the masculine pronoun with reference to the deity except when speaking of the Hindu Brahman which is normally referred to as neuter. I have also tried to alternate between 'he' and 'she' but it may be noted that 'man' is used to include male and female.

In the early days of writing, John Polkinghorne, President of Queen's College, Cambridge and Russell Stannard, Professor of Physics at the Open University, read in rough form some of the scientific material and I would like to thank them. Thanks are also due to my wife who read the manuscript and the proofs and made helpful comments. The book has involved extensive reading in both scientific and religious disciplines and I have tried to acknowledge my debt to various writers in the references but if anyone has been omitted I would apologise in advance. Throughout I have tried to write in non-technical terms for the general reader in order to reach as wide a public as possible.

Robert Crawford

We feel that when all scientific questions have been answered the problems of life remain completely unanswered.

Wittgenstein

1 Where is God?

They seek him here they seek him there
Those theologians seek him everywhere
Is he in heaven or is he in hell … ?

If I ascend up into heaven thou art there,
If I make my bed in hell, behold thou art there.
If I take the wings of the morning, and dwell in the uttermost
parts of the sea;
Even there shall thy hand lead me, and thy right hand shall
hold me …

Ps.139:8–10

The name of this infinite and inexhaustible depth and
ground of all being is God. That depth is what the word God
means.

Paul Tillich

Some, according to our quotations, are looking for God,
others cannot escape him and some are sure that he is
within us. But many today feel that he is some kind of 'absentee landlord' or does not exist. The view may be based
on their experience of life and rejection of the doctrines of
religion or that science has rendered the idea of God obso-
lete. Thus it is said that the Psalmist lived in a pre-scientific
world and it is impossible to accept his belief. It was an age
of blind faith not science which deals in facts. Nowadays
things are explained by natural causes which exclude the
supernatural.

We will examine that claim throughout this book but in the
present chapter we want to consider briefly how the scientific
understanding of the world began to take shape. The ancient
world view was replaced from the 17th century onwards by a
different way of looking at the universe and our place in it and
consequently to effect the religious model of what God was
like and where he might be located.

1

THE ANCIENT VIEW OF THE WORLD

It is sometimes thought that scientific work before the 17th century was of little value, being primitive or culturally relative, but much was discovered. The ancient Egyptians invented a simple method of calculation, the value of pi, divided the circle into 360 degrees, the hour into 60 minutes and a minute into 60 seconds. The Babylonians calculated the length of a year, based on what they thought was the movement of the sun returning to its same position among the stars, and they were aware of eclipses. The Greeks built general laws on things discovered by the Babylonians and Egyptians and left us the schoolchild's nightmare: Euclid's Elements of Geometry!

The Greeks developed mathematics as a valuable tool in the understanding of the universe. Aristotle (384–322 BC) engaged in experiments and Archimedes (287–212 BC) discovered the law of the lever and the principles of hydrostatics, that is, the study of liquids at rest and their pressures. Eratosthenes (284–192 BC) calculated the diameter of the earth and Ptolemy (90–168 AD) produced a summary of Greek astronomical theory. Aristarchus anticipated Copernicus, developing a sun-centred astronomy and both Leucippus (500 BC) and Democritus (420 BC) were forerunners of the atomic theory. Plato (427–347 BC) and Aristotle invented the concept of spheres which they thought carried the sun, moon, planets and stars. These heavenly bodies were perfect and operated in circles, the form of perfection.[1] Aristotle taught that the earth was central and stationary but his philosophical ideas were better than his astronomy. Pythagoras, like the Hindus, believed in the transmigration of the soul while Empedocles and Aristotle had ideas about evolution.

The Arabs after the fall of the Roman Empire preserved the methods of Greek science and translated the works of Greek philosophy and science into Arabic. In many books the claim is made that science rose in the West but Muslim scholars insist that Copernicus, Galileo and even Newton had been anticipated by their scientists. Hence the work of al-Battani (d.929), al-Baruni (d.1048) and ibn al-Haytham (d.1039) is presented. Mathematics and astronomy were studied and logarithms and algebra invented. The Western world was introduced to and adopted Arabic numerals and Abdul Wafa (d.997) developed trigonometry and spherical geometry. Omar Khayyam (d.1123), known for his poetry, solved third and fourth degree equations by intersecting conics which is the highest algebraic achievement of modern mathematics.

Five hundred years before Galileo, al-Baruni discussed the rotation of the earth on its axis and al-Battani measured the circumference of the earth. According to al-Bitruji the orbit of the planet is a sphere or globe which turns on its axis. Ibn al-Haytham wrote on optics and it is alleged by Muslims that his book, *Optical Thesaurus*, is one of the most plagiarised texts in the history of science. Guilty parties include Roger Bacon, da Vinci, Kepler and even Newton![2] Whatever disputes may arise about such claims it is clear that the Arabs were very interested in astronomy with observatories at Maraga, Damascus, al-Rayy, Seville and Samarkand and a similarity has been seen between the planetary models produced at Maraga and Damascus with those of Copernicus. The Muslims made advances in chemistry, botany, zoology, technology, psychology, architecture and medicine.

If the ancient world had made such advances why did significant progress not occur until the 17th century? Some of the reasons were that the Greeks wanted to understand the world and its meaning rather than how it worked: Greek science grew in partnership with philosophy. The method of enquiry was deduction moving from the general to the particular, that is, 'All men are mortal, Socrates is a man, therefore Socrates is mortal'. Mathematics is a deductive science, deducing truths from ideal concepts such as triangles, circles, numbers and symbols. There is no need for observation of nature and, though Aristotle did engage in observation in the source of his biological work, he believed that mathematics was the true science. Observation was a seeking for mathematical forms in the natural order. Hence the circle was a perfect form so the appearance of an elliptical movement must be a mistake in observation.[3] But as we will see it was induction, moving from particular examples to a general truth, that was to prove most fruitful in the development of modern science. It is odd, since the Greeks were so interested in mathematics, that they did not develop numerical notation which would have greatly assisted calculation. It was left to the Muslims to use it, probably having learned the basis from Indian mathematicians.[4]

The Greeks thought that the permanent was the eternal reality and distrusted what we could see and experience in our world of decay and change. Plato, the teacher of Aristotle, spoke of perfect eternal forms or ideals or ideas, present in an invisible world, whereas here are only copies. To reach the highest level of knowledge is not to observe and experiment with the copies but to contemplate the perfect forms of eternal truth. The Indian religious tradition also turns away from an illusory world and seeks, by meditation, the only reality, Brahman or God. Aristotle held that the soul was the form of the body and the faculty which recalls the forms in the world from which we came. The soul is eternal, imprisoned in the body, but we have dim recollections of that other world and true knowledge is recollection of the perfect forms of justice, goodness, beauty, truth. There is a hierarchy of these forms with the Supreme Good at the top. Only a process of education can, according to Plato, reach these abstract and unchanging objects.[5]

Philosophy considered as the operation of critical reason can perceive the eternal forms and measure ethical behaviour by them. When I see an act of goodness or kindness I am seeing a copy of the goodness which is beyond space and time, an appearance of the reality. Underlying the appearances we could discern it, just as the expert can perceive gold and not be deceived by the fool's version of it. Knowledge is recollection and we see it reflected in Wordsworth when he speaks of the child coming from the eternal realm: 'trailing clouds of glory do we come from God who is our home' but the remembrance is dimmed by the taking of the body and the effects of the world: 'soon the prison house begins to close upon the growing boy'.[6] The teacher does not seek to fill the child with information, that is, the empty pot, but tries to get the child to remember what he knows already. Thus, Plato's famous character, Socrates, excelled in asking questions about everything and was accused by the state of stirring up the youth. At his trial he was condemned and offered the hemlock or exile but since he would never leave his beloved Athens he chose the former. He had maintained that the unexamined life is not worth living and this included beliefs about state and religion.

The forms of Plato are eternal, timeless, spaceless and unchanging so that time is an image or copy of eternity. The

Highest Good or Form has been equated with God on the grounds that after the platonic training in mathematics and philosophy there is the need to catch a vision of the Good. Plato's model of God excluded the poetical description of the Greek gods who were immoral. If there was a supreme God he must be the Form of the Good and be pure Being, eternal, self-consistent, unchangeable, and accessible to reason. He must epitomise justice, goodness, beauty, truth, and not be dependent on anything external. Aristotle, the teacher of Alexander the Great, brought the forms down to earth by contending that everything in nature seeks to reach perfection: an ideal state. Every natural process has a last state whose form was present in its first state: in the acorn the oak is present. Unlike Newton, as we will see, who thought of matter as a system of mechanical relations, Aristotle believed it had natural tendencies. A body falls not because it is moved by a force but because it has a natural inclination to fall: the explanation is in terms of purpose.

Aristotle did speak about initial or efficient causes but he was more interested in final causes, what a thing strove to achieve, its end or telos. It is the form which supplies the potentiality to make it what it is to become. Suppose while you are reading this you remember that the kettle is boiling and hasten to make a cup of tea. You ask the scientist who is sharing it with you what has caused the water to boil and he will begin to talk about temperature and molecular structure but Aristotle would say that it is the nature of water when heated to boil. His terminology consists of tendencies, innate properties and potentialities, and he is answering the question why things behave in a certain way, not how. The soul is the form of the body but forms can exist without matter, just as ideas can exist without becoming immanent in some body or project, hence Aristotle contended that there must be pure Form or Being which is the ultimate cause of all motion and all becoming. There is purpose and design since everything is moving towards a goal and this opposed the Epicurean view that the world was the result of random processes.

Aristotle thought of God as the Unmoved Mover. We can think of something which does not move yet influences us. We look at a photograph or painting of a person in our home and remember the ideals which characterised his or her life and feel moved to imitate. Holman Hunt's 'Light of the World' or 'Christ in the Carpenter's Shop' still inspire Christians to follow such an example. Aristotle would have seen the Unmoved Mover in this way luring us towards the ideals of justice, goodness and truth, and yet remaining unmoved and unaffected itself. But the analogy fails in that persons we remember were active in their lives, not unmoved or impersonal. The Aristotelian view in seeking to do justice to the independence and self-sufficiency of God presents an impersonal and static Absolute or Reality which in its influence on religion has posed problems for any faith that sees God as active. How can we do justice in our image of God to One who is the ground of everything yet reveals himself in a personal way? It is a problem which still haunts theology.

Aristotle's cosmology was questioned as science began to advance. The Pythagoreans had contended that the earth was not the centre of the universe but Aristotle rejected this view and his belief was a dominant influence on religion in the medieval age. It was believed that the world was rational and controlled by invisible intelligences, bodies were moved by souls, and falling objects moved naturally towards the earth. Nature had values, purposes and desires but this view was questioned as scientists began to enquire about efficient or natural causes not teleological ones.

THE NEW VIEW OF THE WORLD

Despite the platonic view that the changeable world could not reveal ultimate truth there was a movement to understand the world on the basis of observation, experience and experiment, and it was religion that led the way. Two Franciscan friars at Oxford, John Duns Scotus (1266–1308) and William of Occam (1300–49) began a movement which refused to accept the principles of Aristotle.[7] They insisted that the way to obtain knowledge of the world was by the forming of a hypothesis to explain the facts, and then the testing of the hypothesis by finding whether the facts could be arrived at by using it as a starting point.

In astronomy Nicolaus Copernicus (1473–1543), a Canon of the Church, argued that the earth and planets revolved

around the sun. The idea did not appeal to common sense, so
much of science we will discover does not! When I sit here or
move around I have no feeling that I am hurtling through
space. If I was would I not feel the effect of the wind? But his
view was acceptable to Galileo (1564–1642) and Kepler
(1517–1630). Kepler said that we must think God's thoughts
after him so he recognised that our thoughts must have some
affinity with his. He saw the heavens declaring the glory of
God and God as the great Geometer who filled the universe
with geometrical designs. Much of Galileo's work stemmed
from the invention of the first telescope and with its help he
was able to demonstrate that the celestial bodies were not
perfect. There were mountains on the moon and spots on the
sun hence the traditional view of the perfection of the heavenly in contrast with the earth had to be set aside. The
thought of two worlds – one visible and the other invisible –
gave way to a concentration on the visible world which we
could observe and understand by experiment.

A new paradigm or revolution which removed the centrality
of the earth had taken place. If the earth was no longer the
centre of the universe could God be interested in its inhabitants? The cosmological views of the world presented by the
scripture came under scrutiny since they clashed with what
science was discovering. How could passages which teach that
the sun stood still in order for a leader of Israel to dispatch his
enemies be taken literally (Jos.10:13)? Science was making
theologians think about how the scriptures were to be interpreted but Galileo continued to believe that God was the First
Cause of everything, the initial link in the chain of efficient
causes.[8] He thought that God created everything to be independent and self-sufficient but the active relation of God immanent in the forms or souls was at risk.

The laws of nature could be expressed in mathematical formulae based on what was measurable in nature. Galileo was
interested not in the Aristotelian question: why things behaved
in a certain way, but how they behaved. The thought of
purpose was a hindrance in understanding nature. Initial
causes not final ones came to the fore. What Galileo did was to
measure things, not to look for the evidence of some mystic
form in them: it was size, shape, speed and mass that were important. Such attributes could be weighed and measured in an

objective way and were the primary characteristics of any object. But colour, taste, sound, smell, were subjective or secondary. Reality consisted in matter that could be measured which would exist even if we did not have the secondary qualities. The quantitative aspects had triumphed over the qualitative. Such simple realism sees the world of matter existing apart from our experience of it and true knowledge is the mathematical account of it. The approach raises questions about the reality of our subjective experiences which figure particularly in the arts. But Galileo was not to know that modern science would discover that subjectivity would also enter into physics and precisely with those variables of mass and velocity.[9]

If the new cosmological view presented problems for the literal interpretation of the Bible it also raised difficulties for the Aristotle/Aquinas synthesis. From the beginning of the Christian Church the Greek philosophy had influenced the thinking of the Fathers and it is seen in the work of Thomas Aquinas (1225–74). Aquinas presented a number of arguments for the existence of God relying on Aristotle's stress on purpose. He contended that as every event has a cause the universe must have one and since we see purpose everywhere the world has a Designer. Thus he postulated God as a necessary Being: one whose existence does not depend on anything else, unlike the world and man which are contingent. But if God is a necessary Being how can he create a universe which is contingent? An effect must contain something of the cause. Aristotle realised the implication and contended that the universe was eternal but Aquinas drew back at this point for he remembered the act of creation in Genesis Chapter 1. Later in this book we will look at such arguments but Aquinas is important for his attempt to reconcile reason and revelation. He combined the picture of the Unmoved Mover with the personal Father of the Bible by trying to give a place to the activity of God as saviour. God created the world but also sustains it. He is primary cause but he also works through secondary or natural causes.[10]

Galileo kept stressing that nature not scripture revealed scientific facts. The spiritual knowledge of the Bible concerning the place of man and his redemption could not have been discovered from the world, hence the book of nature and the book of revelation are dealing with two different things. But he did not want to separate religion and science as some scholars do today, for he argued that the latter could also help us to know God. The creation revealed the creator as scripture said (Ps.19; Rom.1:20). Nor did he go to the extent of some modern physicists who maintain that science is a better pathway to God than religion. We accept that science affects religious thinking but religion cannot be based on science for a change in the scientific paradigm might destroy such a construct. Neither did he think that science can explain everything and we have no need of the God hypothesis. Natural theology has been much discussed in our century, but it is generally accepted not to be on the same par as that given through the prophets, mystics, saints and founders of the various religions.[11]

Science proceeded with Francis Bacon (1561–1626) who placed so much stress on the experimental method that at the age of 65, having exposed himself to extreme weather conditions in an effort to try to stuff a chicken with snow, he caught flu and died. He wanted to know how long it could be preserved. Bacon neglected theory and creative imagination in arriving at knowledge of the world, for he believed 'that science consists of the accumulation and classification of observations. He insisted that induction is the easy road to knowledge; make observations, summarise them and generalise.'[12] But we will see later that observations are theory laden and there is selection according to interest.

The stage was now set for the model of the world as a machine. Isaac Newton (1642–1727), reckoned to be England's greatest scientist, proceeded to define the laws of mass, position and velocity and taught that matter was composed of solid and impenetrable particles. Scientific concepts were literal representations of the world and forces and masses replaced purposes. Newton's work meant that it was possible to predict when eclipses would occur, the appearance of comets, and to know the exact position of the stars and planets both in the past and future. Matter may be rearranged but there is no basic change.

Everything is in motion. A mass left undisturbed moves in a straight line without any change of speed. The force that acts on objects is gravity so that the moon is influenced by the pull of the earth and it in turn affects the ocean tides. The mass of anything is not its size but the amount of matter which is in it and such a measure will affect the pull of gravity between it and another body. The greater the mass the greater the attraction. The further apart bodies are the weaker the pull of gravity between them. The world is God's machine for he made it and maintains its laws.[13] The laws are the expression of his will and they reveal that the world has been designed by him. Space and time are absolute and separate. But if God is absolute and infinite how can space and time also be absolute? Newton was aware of the difficulty and suggested that both are attributes of God and he penetrates them. Hence he could contend that God is both immanent as well as transcendent but most commentators insist that his science leads to a deistic God who adopts a spectator-like attitude to the world. But he did say that space was the sensorium of God and he could perceive all things in it. The question is, and it is a debatable one, whether such perception means the immanence of God in all things.

We can think of Newton's view as the container or box-like model of space. It contains all things. Every object has a definite place and God holds the box together. Thus he sustains the universe and everything in it. This looks like the Hindu view of Brahman who is the holy power implicit in the universe or the ultimate reality underlying it, and as the Absolute, contains all things. It would be pantheism, God is all and all is God, merging all things in the divine and denying his personality. But Newton tries to avoid such a conclusion by speaking of God as the Lord and Governor of all things and refusing to accept that he is the soul of the world. However, one main focus of the immanence of God is incarnation and Newton could not accept this because of his container view of space. A container that holds water cannot become part of the water, hence God cannot become man. His theological position, then, was unitarian which posed difficulties in obtaining a professorship at Cambridge for it required subscription to Anglican belief. His views about God led to deism, that is, belief in the existence of the deity but not in revealed religion or in a personal providence. However, Newton believed in a designer God who had made the machine and in his work, *Optics*, eulogises the beauty and order of the world:

How came the bodies of animals to be contrived with so much art, and for what ends were their several parts? Was the eye contrived without skill in optics? ... Does it not appear from phenomena that there is a Being incorporeal, living, intelligent ... [14]

The example of the eye is significant for it was something that presented problems for Darwin and his followers in arguing against divine design. Newton could not accept that the arranging of forces in the world could have happened by chance, hence the law of gravity expressed the will of God. Most of the scientists who founded the Royal Society for the advancement of science in the latter half of the 17th century were religious and followed him in that belief, but they thought that the purpose of God was external to the world, not immanent. Hence the Aristotelian view was lost: God was First Cause not Final. If God intervened in the world it was a rare occasion and he only did so in order to correct irregularities or by way of adjustment: a celestial plumber! But the pattern of the planets could not be explained without him. The French scientist, Pierre Laplace (1749–1827) however, was able to account for the irregularities without the aid of God and his work showed the inadequacy of basing theology on 'the God of the gaps', that is, God introduced to explain what current science cannot. The God hypothesis could be disposed of. The only exception to a mechanical world was man who had the reason to understand it and could tune into it, 'because the rational order of nature is akin to human reason'.[15] But if our minds are in tune with the universe would that not be a pointer to a rational Mind behind it?

In summary, Newton thought he was describing the world as it was, not simply how it behaved, and that once we knew the position and speed of matter we could predict with accuracy its future state. Nature was deterministic rather than teleological. The scientist was independent of what he studied with events located in space and time considered as an absolute framework. Time passed uniformly and universally and was the same for all observers. Every object has a definite place in space and it is separate from time. Mass and velocity could be expressed in mathematical formulae and were objective characteristics of the world independent of the observer. It was assumed, in agreement with common sense, that the length and mass of an object are unchanging, intrinsic, objective properties and we had no effect on them. Such simple realism will be challenged in subsequent chapters.

Newton excluded mind from the mechanical view of nature for it was the key to our understanding of the world and the distinguishing mark of mankind. Rationality became the key word of the century that followed and was aptly called: 'The age of Reason'. Newton accepted the view of the father of western philosophy, Descartes (1596–1650), who separated mind and matter: dualism. Descartes thought of the world and our bodies as machines but mind was not in that category for it directed their operation. He was a mathematician and argued from a priori truths, that is, prior to experience. Like Plato he believed that experience can only tell us something about the 'appearances' of the world for the real is the rational. Again, there is a turning inwards, not only to rationality but also intuition, since there are clear and distinct ideas which are self-evident. One, as we will see later, was the idea of God which is innate. There is change in the Newtonian world but only in the sense of the rearrangement of matter whose form is fixed. The approach was also reductionist since it was thought that what was happening at the lowest level – physical mechanisms and laws – was determining events, apart from those in the human mind.[16]

This was the new picture of what the world was like and the relation of man and God to it. Meaning, value and purpose which had been present in the old view now became subservient to a mechanical world. Form in matter was the important essential for the Greeks but Descartes eliminated it. His stress on the mind resembled the idealism of Plato and Aristotle and he emphasised the idea of God in humanity when he tried to prove his existence. Aristotle's view of God as pure thought may not have been far away from his understanding. But the centrality of the earth had disappeared and the focus was on this world not some invisible platonic or religious one. The perfection of the latter had been queried by Galileo's telescope. In consequence, a dualism of body and soul had emerged. The goal of the old view was to glorify God and enjoy him forever but the new view with its concentration on initial causes stressed the mechanism of man's nature. How he related to the world became more important than any purpose with regard to a creator. But it was the mechanical process of natural selection that we will consider in subsequent chapters that was to deal the old view an even more deadly blow. A look back at 16th century European society shows it was dominated by religion, both Christian and Muslim, with God present in all things. Pilgrimages, rituals, belief in spirits and relics, magic and astrology, predominated in popular culture and the authority of religion was taken for granted. But the new view of the 17th century placed God in external relation with the world and mankind so the problem was how to recover his immanence in all of life.

The physics of Newton dominated from the middle of the 17th century to the mid-19th but he had left certain problems unresolved. One was connected with light. He thought that it was composed of particles but this was challenged by a Dutch scientist, Christiaan Huyghens (1629–95), who argued that light consisted of waves. He was supported by Thomas Young in the following century but it was in the 19th that Clerk Maxwell demonstrated the wave theory of light and showed that the waves were oscillations of coupled electric and magnetic fields. Today we have the paradoxical situation of understanding light as behaving either as a wave or particle depending on circumstances.

It was thought by many that Newton had said the last word:

Newton and Nature's laws lay hid in night,
God said, 'Let Newton be'
And all was light.

But Alexander Pope had been too optimistic for:

Nature and Nature's law lay hid in night.
God said, 'Let Newton be' and all was light.
It did not last; the Devil howling 'Ho!
Let Einstein be!' restored the status quo.

Einstein paid tribute to Newton:

Let no one suppose, however, that the mighty work of Newton can easily be superseded by relativity or any other theory. His great and lucid ideas will retain their unique significance for all time as the foundation of our whole modern conceptual structure in the sphere of natural philosophy.[17]

Science advanced not only in physics but also in medicine with William Harvey (1578–1657) and Andreas Vesalius (1514–64); and in chemistry with A. N. Lavoisier (1743–94). Joseph Priestly (1733–1804) discovered oxygen; and A. N. Linnaeus (1707–78) and Georges Buffon (1707–88) made progress in biology. But there were forces other than science that were changing the Christian world view: humanism, intellectual authority and secularisation. The importance of reason, the transition from an oral to a print culture, the questioning of religious authority, the 'weighing' and 'measuring' of evidence – all contributed to the changed outlook. People slowly began to think that just as the scientist could control his experiments they could manage their own destiny. Beginning in the 16th century both culture and belief began to undergo reappraisal. Man might indeed be the measure of all that he surveyed.

The unity of Christianity had been disrupted by the Protestant Reformation but there was also the power of Islam which in the early 16th century had enormous territory stretching from Hungary to the Persian Gulf. Constantinople, formerly the centre of Greek Orthodox Christianity, fell to Ottoman cannon in 1453 and became Istanbul, the largest city in Europe with its beautiful mosques, palaces and great bazaars. As the Muslim conquest continued panic spread throughout Western Europe and the pope insisted that the Ottoman advance was a punishment from God. Culturally the West had to re-examine its beliefs and understanding of society as it made contact with other religions. The experience of Portuguese and Spanish travellers among the societies of India, East Africa and South-east Asia encouraged the belief in a religious pluralism. They observed Muslims, Buddhists, Hindus and local religions living together in a more tolerant way than in Europe. In 15th century Spain there were 100,000 Jews and 700,000 Muslims. The Jews were attacked and deported and crusades mounted against the Moors.[18]

Both the Jewish and Muslim religions challenged the beliefs of Christianity in a way that was more radical than the divisions within Christianity. We think of our century as raising the problem of religious pluralism but the seeds of the debate go back much further. However, the religions agreed that there was a spiritual presence in the world but it was to be challenged by science.

2 Is God in the World?

In my most extreme fluctuations I have never been an atheist in the sense of denying the existence of God.

Charles Darwin

The centre of interest shifted in the 19th century to the question of the origins of mankind. Biology focused on the evolution of humanity which raised questions about our nature, status, and relation to God. Are we animals superior to subhuman nature but without a soul or special kind of essence? Can moral behaviour be explained without God? We will begin to think about these questions in this chapter and continue to probe them in later ones.

With regard to the concept of God, deism made him remote, and it provoked a reaction in the romantic movement. Poets such as Wordsworth spoke of feeling a presence that disturbed with 'the joy of elevated thoughts, a sense sublime of something far more deeply interfused'. It is in nature and the mind of mankind, a kind of spiritual energy which 'impels' all things by its mobility. He was influenced by Greek Stoicism which taught that the world soul or spirit is related to the world as the soul is to the body and penetrates all things. The spirit is a refined form of matter. The importance of feeling emerged also in the religious revival of the 18th century with John Wesley when he felt his heart strangely warmed by a conversion experience. But Darwin's evolutionary theory implied the development of life forms in a mechanistic and naturalistic way without the need for any divine guidance.

We shall look at Darwin's home background and education briefly before considering the theory. Charles Darwin (1809–82) did not show in his early years any signs of genius. In his *autobiography*, a model of clarity and simplicity, he admits that he was considered both by his father and teachers to be a very ordinary boy below the common standard of intellect.[1] It was his father's intention that his son should be a doctor but Charles was not suited to it and decided to become a clergyman. He was 19 years of age when he went up to Cambridge to study for the Church and he had little doubt about the strict and literal truth of every word in the Bible or the creed of the Church of England. He showed some ability in the classics but was inept at mathematics. What impressed him most were the writings of William Paley whose *Evidences of Christianity* and *Moral Philosophy* presented the arguments for the existence of a designer God. In his final examination for the BA degree he did not take honours and was placed tenth in the list of the pass degree candidates. It was clear that his abilities were more in the natural sciences.

Darwin had always been interested in and a keen observer of nature. Beetles in particular were collected and studied and most of his reading was in natural science. Then an unexpected offer occurred to travel as a naturalist on HMS *Beagle* and after some doubt and dispute with his father, Darwin sailed on 27 December 1831. He returned to England five years later on 2 October 1836. His interest in geology had been fostered by Charles Lyell's *Principles of Geology* which asserted that the history of the earth was due to the accumulative action of natural forces rather than major catastrophic upheavals such as the story of the flood in the Book of Genesis. Darwin was to apply this slow cumulative process to the evolution of life. For two years during the voyage Darwin maintained his orthodox belief in Christianity and he records being laughed at by several of the officers for quoting the Bible as an unanswerable authority on some point of morality.[2] He also appreciated the work of missionaries in civilising the natives of the lands that he visited and wrote home arguing that more missions were needed.[3] Nature with all its grandeur inspired religious feelings: 'No one can stand unmoved in these solitudes, without feeling that there is more in man than the mere breath of his body.'[4]

But reasons against belief crowded into his mind. He gradually came to doubt the authority of the Old Testament because of its false history (the Tower of Babel, the rainbow as a sign), and its attributing the feelings of a revengeful tyrant to God. These accounts he felt were no more trustworthy than the sacred books of Hinduism! Darwin appeared to understand the biblical stories literally but he must have been aware that the scripture is full of allegory, myth, metaphor and symbol, and that the Fathers of the Church had recognised it. He did not resolve his doubt by accepting that in scripture there is a developing idea of God from more primitive images to the high moral God of the eighth century prophets. But he had problems with miracles which opposed the fixed laws of nature, the ignorance and credulity of the people who recorded them, the fact that the Gospels were not written at the time of the events and differed in important details, and the doctrine of eternal damnation. Since he was raising theological questions before the evolutionary theory took shape in his mind his disbelief cannot simply be accounted for on that basis.

The theory that nature selects those that will survive demolished his belief in Paley's argument for design, for everything in nature follows fixed laws.[5] There seems to be no more design in the variability of organic beings and in the action of natural selection, than in the course which the wind blows. Everything develops naturally and nature was 'red, in tooth and claw'. How could an omnipotent and omniscient God allow the sufferings of millions not only in this world but in that which is to come? But disbelief is accepted with reluctance as he returns to the argument for design:

> Another source of conviction in the existence of God, connected with the reason and not with the feelings, impresses me as having much more weight. This follows from the extreme difficulty or rather impossibility of conceiving this immense and wonderful universe, including man with his capacity of looking far backwards and far into futurity, as the result of blind chance or necessity. When thus reflecting I feel compelled to look to a First cause having an intelligent mind in degree analogous to that of man; and I deserve to be called a Theist.[6]

We will see that the design argument today puts this stress on the whole rather than the parts of the process. The belief was strong in Darwin's mind when he wrote the *Origin of Species* (1859). In the final paragraph of the second edition he wrote:

> There is grandeur in this view of life, with its several powers, having been originally breathed by the Creator into a few forms or into one; and that while this planet has gone

cycling on according to the fixed law of gravity, from so simple a beginning endless forms most beautiful and most wonderful have been, and are being, evolved.[7]

But the conclusion grew weaker because it was based on the lowly human mind developed from that of an animal. How could it be trusted when it draws such grand conclusions? It is, as Darwin said, like a dog speculating on the mind of Newton. He is recognising the limitations of reason which some of his followers today fail to do. Darwin concludes that he is unable to explain such problems and must remain an agnostic. He regretted truckling to public opinion in using the term 'creator' rather than some wholly unknown process and in his *Descent of Man* published in 1871 he writes that there is no evidence that man was endowed with the belief in God's existence. Some races have no idea of one or more gods though there seems to be a general belief in a creator and in unseen or spiritual agencies. The Fuegians whom Darwin was familiar with had such beliefs but did not believe in what he would have called a god nor did they practise any religious rites.[8]

THE THEORY

Darwin noted in the various places which he visited on his voyage the savagery of nature, the extinction of many species, the wide gulf between the savage and civilised person, and the variations which continually presented themselves. He was depressed by the sight of the Fuegians with no proper clothes, no fit language, and no decent homes or property, except bows and arrows. He reflected that there was a tremendous difference between civilised and savage man, greater than between a wild and a domesticated animal. Yet they were fitted for their environment and a different one would not have been suitable. One of the Fuegians had been taken to England and brought back aboard the *Beagle* to encourage civilised ways among his tribe but when left for a while with them he returned to their customs and way of life. Darwin reflected that no civilising influence could change his instincts and the power of his home environment and given the chance to return again to England he refused.[9]

In a close study of nature Darwin began to think that species might not be fixed and unchangeable. One example was a group of small birds called finches which had about 14 different species that differed in beaks, colour, feeding habits and size, but were closely related. The relationship showed a common ancestral species which had developed into a new species, hence Darwin concluded that species were not immutable. On his return home he spent a lot of time thinking about the domestic breeding of animals and consulted with skilful breeders and gardeners and realised that selection was the keystone in making useful races of animals and plants. But how could selection be applied to organisms living in a state of nature and could natural selection produce new species?

The museums of the world are full of collectors' items but it takes a genius to make connections and establish links so as to form a theory. Darwin spent ten years classifying his materials and thinking about breeding but a theory only emerged by accident when he happened to read the Revd Thomas Robert Malthus' *Essay on the Principles of Population* (1798). Malthus (1766–1834) pointed out that the population increased more rapidly than food supply and only checks such as poverty, plague, famine and war prevented community disaster. He advocated the postponement of marriage and procreation but being a clergyman was wary about suggesting birth control. It was a natural law that we must struggle to produce food in order to survive and with an ever increasing population the struggle becomes more fierce but without this stimulus we would never have emerged from the savage state.

Darwin, reading Malthus simply for amusement, suddenly realised the parallel with the struggle for existence which he had observed in nature. It occurred to him that favourable variations – height, strength, speed, cunning – would be preserved in such a struggle and the unfavourable eliminated. Once a species becomes stronger than another it will dominate and will be able to adapt to a different environment and evolve into another species. Further there are organs in different species which resemble one another. If we are out in a small boat for amusement we use the hand as a paddle and are thus imitating the porpoise paddle by which it pushes itself through the water. It has similar bones to that found in the human hand and these are also in the leg of a horse and the wing of a bat! Again, the embryos of mammals, birds, lizards and snakes in their early stages are very similar to one another. Darwin told a story about von Baer, the founder of embryology, who preserved two embryos in pure alcohol and forgot to label them; afterwards he was unable to tell whether they were lizards, birds or mammals. A fish breathes through its gills and the embryos of a man and dog show gill holes in the neck which correspond to the gills of the fish. The human embryo also shows the beginning of a tail.[10]

Darwin made a brief abstract of his theory but one problem puzzled him: the tendency in organic beings descended from the same stock to diverge in character as they become modified. Thus we can class species of all kinds under genera, genera under families, families under sub-orders, and so forth. He only overcame this problem by a flash of insight or intuition which he ruled out in religion! The solution occurred to him while sitting in his carriage that the modified offspring of all dominant and increasing forms tend to become adapted to many and highly diversified places in the economy of nature. In other words, in a changing environment those who can adapt best will survive and gradually a new species will emerge.

The survival of the fittest in a context where more organisms are produced than can survive, inheritance and the ability to adapt became the key factors in evolution with variations which confer an advantage leading to a new species. Darwin clarified what he meant by the latter: 'I look at the term species, as one arbitrarily given ... to a set of individuals closely resembling each other, and that does not essentially differ from the term variety, which is given to less distinct and more fluctuating forms.'[11]

The definition is based on comparison of features of organisms and differs from what is acceptable today where reproductive isolation is the criterion for separating species. But he found the question of inheritance a problem. Variation does occur within a species or family likeness and offspring often inherit the characteristics of their parents. These variations are helpful because they can enable an organism to survive and they are passed on from generation to generation. Reproduction becomes important as does the inheritance of features upon which natural selection can act. Organisms which are more successful by reason of some variant leave more offspring than those without variant, thus there is natural selection or survival of the fittest. The modern emphasis is more on this reproductive element than in Darwin who tended to see 'fitness' as some feature possessed by an organism which enabled it to adapt and increased survival chances, for example, birds possessing wings.[12] The adaptation did not mean purpose but was a response to the environment.[13]

It has been difficult to reach agreement on how species do arise. Some argue that natural selection is only one of a number of processes that bring about evolutionary change. Does natural selection act upon the genes which are the units of selection and determine what we are or should we stress the whole organism and the influence of the environment? Is evolution more connected with the group or the species as a whole? We will return to this in the next chapter. Darwin dealt with the origin of man in his *The Descent of Man*. He thought that gradual modification would have taken place in anthropoid ancestors so that the upright posture and larger brain developed. Man's moral qualities and his mind were derived through natural selection from lower forms. On a visit to the zoo in London he was impressed with the behaviour of Jenny the orang-utan: her intelligence, affection, passion, rage and sulkiness. Comparing her with the savage who roasted his parent and was naked and artless, he pointed out that Jenny did not suffer by comparison![14]

Yet Darwin did admit that the difference between the mind of the lowest man and that of the highest animal is immense. The ape could plunder, use stones for fighting or breaking nuts but could not fashion a stone into a tool. He could not reason or reflect on God and was not capable of disinterested love. Man is self-conscious, can engage in abstract thought, and has a highly developed language but the difference between man and the higher animals, great as it is, was one of degree and not of kind. Many of our characteristics can be seen in the lower animals: love, memory, attention, curiosity, imitation and signs of intelligence, and they are capable of improvement.[15]

With regard to the uncivilised humans one thing did impress Darwin and that was their power of language imitation. He was amazed in the light of the European difficulty of distinguishing apart the sounds of a foreign language how easily the Fuegians mimicked correctly what he said and remembered it. Is it, he asks, due to the more practised habits of perception and keener senses, common to people in a savage state as compared with the civilised? But with respect to the moral and mental aspects he believed that they were part of evolution and connected with the social instincts. However, he did state that the moral sense revealed the best and highest distinction between the human and the lower animals.[16] But such an immense difference of degree could be interpreted as a difference in kind.

It was also argued that Darwin was giving natural selection too large a role in evolution and in his *Descent of Man* he admitted that this was the case. Further, in the working out of his theory Darwin had paid special attention to domestic breeding which did show the great variety that could be produced but did not demonstrate that a new species had originated. And, if the mind of man was required in producing and controlling such variety, why could it not be maintained that a cosmic Mind lay behind the evolutionary process?

DESIGN

It was widely held before Darwin published the *Origin of Species* that a Creator had designed creatures so that they would adapt to their environment. His purpose was that they might benefit from living in such a world and they might glorify him in what they did.

Darwin himself was impressed by the wonderful way that animals and plants were adapted to their environment which seemed to indicate design but he explained adaptation by natural selection of chance variations conferring advantage in the struggle for existence. Some organisms were better fitted to survive than others.[17] The explanation conflicted with both Aristotle and Paley, the former postulated an internal form or soul or principle directing the organism and the latter contended for external design. Darwin explained adaptations not by design for a particular purpose but as conferring an advantage.[18] Birds did not evolve wings in order to fly, that is, purpose, but by enabling them to fly the wings were a major contribution to survival. Natural selection replaces God as a cause of adaptations which confer an advantage. Yet in such a rejection Darwin is not sure, as he said in his letter of 1870 to J. D. Hooker:

> Your conclusion that all speculation about preordination is idle waste of time is the only wise one; but how difficult it is not to speculate. My theology is a simple muddle; I cannot look at the universe as the result of blind chance, yet I can see no evidence or beneficent design, or indeed of design of any kind, in the details.[19]

We see again his point of applying design to the whole. A Designer could use the process of evolution and ensure that the conditions were right for life to start without interfering with the course of development. 'It is in every case more conformable with what we know of the government of this earth, that the Creator should have imposed only general laws' than of one who 'kept fiddling with nature every time a new species is to be created'.[20] The model of God is the framer of the laws of evolution. At least this appears to have been his attitude up to 1850. Darwin had the problem, which has been evident ever since, that evolution cannot simply fly off in any direction for some adaptive options are more likely than others. Functional adequacy is all that we can expect from natural selection and competition will improve the adaptations.

　In the modern view, fitness to survive depends on the ability to leave progeny which can themselves reproduce: to propagate genes in subsequent generations. But how do the adaptations arise in the first place? There may be something already there in the organism that can be put to some other use and thereby confer an advantage. For example, the bones of jaws were already present in fish but they were being used for breathing not for eating. If reptiles had not possessed scales would the feathers and wings of birds have ever evolved? It is very tempting to say that there is some kind of purpose going on but evolutionists refuse to admit teleology and contend that it relies on chance adaptions of new functions. It was the role of chance in the process that caused problems for religion for it seemed to exclude order and design. Darwin had concluded that odd variants might occur by chance since nature juggled with the variations discarding some and accepting others.

He argued against design in 1868 when he wrote that if God is responsible for everything then the bad as well as the good variations are due to him. Darwin was convinced that organisms 'during an almost infinite lapse of time' have had their organisation rendered plastic and variations both good and bad occur. Things make themselves, but if this is so then God is not responsible either for the good or bad variations. Asa Gray tried to persuade Darwin that God was guiding the good but Darwin believed that God determined everything so he is responsible for the bad as well. This questioned his morality.[21] We will put forward reasons later for modifying this image of God in a discussion of the divine attributes.

Darwin did have problems about details, particularly the human eye, and admitted in *The Origin of Species* that if it could be demonstrated that any complex organ existed which could not possibly have been formed by numerous, successive, slight modifications, his theory would break down. But modern Darwinians argue that they do not know of a single case of a complex organ that could not have been formed by numerous successive slight modifications.[22] The objectors retort that it may be possible but it is not probable. First, the existence of a continuous sequence is necessary; secondly, that it can be shown by quantitative estimates that the immense number of mutational steps could have occurred; and thirdly that it could have happened in the time available. Since the whole process is random and natural selection must have been able to trace out the best adaptations, it is asserted to be improbable.[23] Natural selection demands that a character of an organism should bring an advantage at every step in its evolutionary history but what use is a half-evolved eye?

However, the old teleological arguments based on nature as a finished product had received a death blow. Nature was now seen as developing, not a finished product; the machine model had given way to the organic. Order, it was argued, could be developed spontaneously and teleology was regarded as a hindrance in understanding how an organism functioned. But T. H. Huxley, 'Darwin's bulldog', admitted that the mechanist or reductionist has to assume an initial molecular arrangement in order for the things in the universe to evolve and therefore he is always at the mercy of the teleologist who asks him to disprove that such an arrangement was not intended.[24] The design argument for the existence of God returns in that shape as we will see in a later chapter.

It was also argued that Darwin had placed too much stress on the conflict observable in nature and not enough on cooperation and altruism. There is a social unity of animals without which they would not survive and this opens up the question of group behaviour. When, for example, a bird gives the alarm at the approach of the hawk it is indicating its position and could be caught. How can such altruism be explained by natural selection, for the bird has displayed a characteristic which though advantageous to the group is a disadvantage to itself? As a consequence it is contended that group selection is the unit and not the individual: the sacrifice of one favours the many who are able to escape and increase their numbers. The argument rests on altruism being widespread in the group and an isolation from selfish groups who by interbreeding might contaminate them. Altruistic behaviour is based on kinship or the parental instinct or that experience teaches the bird that it can confuse the hawk with its call. Therefore fitness is redefined as an inclusive term which favours both the individual and his kin but whether the altruistic behaviour is genetically determined or learned is another question.

Darwin considered group selection because he had to account for the development of the moral faculties. Animals, he pointed out, do realise the value of social behaviour because they need aid when under attack and do not like to be isolated from the group. There was sympathy, fidelity and courage among animals and mankind, but he had to counter the objection that animals who sacrificed themselves for others would not leave offspring so altruism would not pay in the struggle for survival. His reply was that tribes endowed with such qualities would be able to defend themselves better than those where selfish traits predominated. Sacrifice would be praised and selfishness condemned in tribes that were courageous and fought for one another despite the loss of individual members, hence he concludes that tribes with high moral qualities would be victorious over most other tribes.[25]

It is recognised today that chance and necessity play a role in evolution: change occurs through random variations but they are controlled by the law of the advantage which they add to the organism. The law could reflect design for it is difficult to speak of laws being blind as Darwin did. Chance can be seen as releasing nature from determinism and reflecting the spontaneity, creativity and novelty which we observe in nature. But the difference about the nature of mankind in Darwinism and the Judaeo-Christian tradition was crystallised in the debate about species. In the latter, man possessed a mind or soul that made him different from the animals. A. R. Wallace, who arrived at the theory of evolution at the same time as Darwin, argued for the distinction. He refused to accept that there was only a difference in degree between humanity and the animals and contended that while the body was developed 'by the continuous modification of some ancestral animal form, some different agency, analogous to that which first produced organic life, and then originated consciousness, came into play in order to develop the higher intellectual and spiritual nature of man.'[26] He proceeded to mount the argument for the distinctiveness of man by contending that there was a much larger gap in intellect between man and ape than Darwin had acknowledged and primitive tribes could not fill the gap for their inherent mental capacities were the same as civilised peoples. Moreover, Darwin was wrong in seeing little difference between animal signals and human speech and that morality and conscience had evolved through the social instincts of mankind.

How is Wallace's view to be explained? Some have said that he exhibited cowardice in not being willing to follow through his belief in natural selection or being unable to escape the belief in human uniqueness. But Stephen Gould believes that it was Wallace's belief in natural selection which convinced him.[27] Wallace was one of the few non-racists of the 19th century in that he contended that the brains of 'savages' are neither much smaller nor more poorly organised than our own. They can be introduced to culture and civilised and come to use their brains more extensively than in their primitive state. But natural selection can only fashion a feature for immediate use. The brain is vastly overdesigned for primitive society hence natural selection could not have built it. If the capacities that we have originate because of a future need it is possible that a designer could have given the process that direction.[28]

There is an ethical tension in Darwin for he wants to retain the Christian ethics of love and compassion yet he knows this would lessen the competitive struggle and undermine what he had understood to be the source of progress. The crux of the matter is that ethical norms cannot be derived from evolution as the failure of Herbert Spencer's social Darwinism shows. Huxley said that what is ethically best opposes 'the gladiatorial theory of existence.'[29] Huxley believed that a sort of moral intuitionism was the source of ethics.

Belief in a deity is not innate according to Darwin for if it was then we would be compelled to recognise the existence of cruel spirits which are also believed in by primitive people. It is the advance of civilisation that makes man believe in a more benevolent deity. The argument did not satisfy Wallace for he was pessimistic about the moral progress of mankind but Darwin retained his optimism about man who in his bodily frame bears 'the indelible stamp of his lowly origin' but has 'noble qualities, sympathy, benevolence and god-like intellect'.[30] The demise of Darwin's belief in Christianity culminated with the death of his little ten year old girl, Annie. He could not understand how this could have happened to the one who was the joy of his life. It destroyed the 'tatters of belief in a moral and just universe'. He said later that this period chimed the final death-knell for his Christianity, even if it had been a long, drawn-out process of decay. 'Charles now took his stand as an unbeliever.'[31] He had thoughts of suicide and was worried about his declining mental powers.

Darwin died in 1882 and was buried in Westminster Abbey despite his statement: 'I think that generally (& more & more as I grow older), but not always, that an agnostic would be the most correct description of my state of mind.'[32] In a recent book this agnosticism has been denied in favour of atheism,[33] but in the funeral oration Dean Farrer said that Darwin was not a materialist but a 'healthy, noble, well balanced wonder of a spirit, profoundly reverent, kindled with deepest admiration for the works of God'.[34] As to why he was buried in the Abbey there is considerable speculation, with some seeing it as the Trojan Horse of naturalism entering the fortress of the Church! Suspicions have been raised since it was the desire of Huxley and other members of what was called the X club for the burial to take place in the Abbey.

Why did they insist on this Christian burial place when it was the desire of Darwin's widow for a local resting place? The club represented scientists dedicated to liberating science from theology and encouraging an intellectual priesthood. Darwin was their high priest and for them his burial in the Abbey was symbolic of the secular entering into the most sacred place of Anglicanism.

RELIGION

It is difficult for us to realise what a shock the theory of evolution had for the Victorian age. The picture of the world had changed from a machine which had a Designer to an evolving organism based on blind chance. The Victorians with their class consciousness were disgusted by the thought that their ancestors had emerged from the jungle. Mankind had been regarded as special since it had been created in the image of God and received a living soul but now it appeared the mind or soul had emerged in the process of evolution. God's relation with humanity was under threat for where was he in the process? How could they continue to sing the hymns that praised the Almighty Creator? The doctrine of the Fall of mankind showed the need for a Redeemer to come to the rescue but evolution moved from lowly origins to the perfection of humanity. It reversed what Christianity said: man being created perfect and then 'falling' into sin. Genesis taken literally did not agree with Darwin and if it was wrong about creation, could other parts of the Bible be trustworthy? Christian morality was based on doctrine, if the latter was suspect would the former not be discarded? Victorian values were grounded in Christianity but if it was shaken what would happen to the fabric of society? The whole scenario demonstrates that religious systems have elements that are closely knit and if one is falsifiable the whole edifice could come tumbling down.

Some theologians rejected the theory, others tried to accommodate it into the Christian framework, and some lost their faith. Those who did try to accommodate were often accused by conservatives that they were not being honest in changing the Faith and by atheists that they were disguising the truth. But the scientific community was also divided in its response to religion. There was a conflict between science and religion but it was with some versions of Christianity rather than others. There were extreme positions on both sides with John Tyndall making the claim that science could explain everything and W. K. Clifford denying the distinction between body and mind. Some scientists believed that religion should be preserved but its roots were in the depths of man's nature. Darwin was willing to think of God impressing laws on matter in such a way that 'the production and extinction of the past and present inhabitants of the world' were due to secondary causes[35] and T. H. Huxley believed that science had nothing to say against a God who had set the process going.

We have already noted Darwin's tendency to interpret the Bible literally, and this creates problems. According to the Bible chronology creation occurred in 4004 BC (Ussher) whereas science dates the Big Bang about 15,000 million years ago, though the time span was not known in Darwin's time. There is the problem raised by physics of how could we have a first, second and third day without a sun and moon and stars? But literal interpretation is not necessary for images of creation and the divine point beyond themselves. Who thinks literally when the scripture says that God is a rock, or shepherd or father or husbandman or … ?

In Genesis there are two stories of creation. Genesis 1 states that the earth was created first and then the sun, and places the origin of birds before reptiles. It is incorrect scientifically but we do see the insight of the biblical writer when he says that life originated in the sea and creation moved upwards to man. But each is created according to his kind and man is distinguished by being made in the image or likeness of God (1.20:26). Image has been interpreted in different ways: righteousness, holiness (Eph.4:24), ruler (Ps.8:5:8), knowledge (Col.3:10), relationship and openness to God. But distinct species opposes the Darwinian account. In the second story of creation, which is generally regarded as less sophisticated, the order of creation is wrong, as it begins with man and works downward (2:7) but it has the insight that man was created from the dust of the ground. Since the theory of evolution in its modern form envisages life evolving from inanimate chemicals there is an agreement about lowly origin. The Hebrews saw clearly that there was something wrong with mankind: it was an empirical fact. How could they explain the evil in the world and humanity if God had created a good cosmos? God could not be the author of evil? Hence these stories seek to explain how a good creation became bad. It is their explanation in vivid dramatic form.

The stories of creation contend that the relationship with the Creator was broken because of a misuse of free-will. Robots were not created, rather free beings who would respond to the love displayed by the creator. The creature, however, desired full autonomy and wanted the knowledge possessed by God so that he would be a god (Gen.3.5). Sin seems to be loss of ignorance and the desire after knowledge, power and independence from God. The stories are not based on eye witnesses, how could they be? Hence they are not historically or scientifically true but there is psychological truth.[36] The story of the temptation of Eve and how she fell has been repeated over and over again. She looked at the forbidden fruit and kept looking! The look led to longing and the longing led to laying hold. Constant looking and thinking creates the desire to possess (3.6). Thus it is repeated in the case of David with Bathsheba and stands in contrast with Joseph and the wife of Pharaoh. He did not dally with the temptation but fled from it!

Some thinkers in the Victorian age felt that they could dispense with the ancient cosmology and record of the creation and simply say that the account was stressing man's relationship to God and the dependence of the world on him. Others saw that the Fall of Adam and Eve was symbolic of the fall of everyman: 'each man is the Adam of his own soul.' Many were on the way to understanding the Bible in a non-literalist fashion, as confirmed by *Essays and Reviews* (1860) where the findings of geology were accepted. Revelation, it was argued, was not confined to Christianity, and it was necessary to examine the relation of revelation and reason. There was a reference to the self-evolving powers of nature, scripture and inward experience must be able to stand up to critical examination, and there was a need to understand in the light of science how God is related to the world. Uniformity of doctrine and belief were not possible, and the miraculous cannot be taken for granted.

Historical criticism which raised questions about authorship, inspiration, eternal punishment, and so on, was acceptable and the Bible must be interpreted like any other book. Doctrines and creeds based on external evidence gave way to basing religion on religious feeling and experience and they in turn became the criterion of beliefs which did not correspond to moral and religious intuition. Much controversy ensued between Liberals and Evangelicals over the publication. However, by the end of the century when a younger group of Anglo-Catholics published *Lux Mundi* (Light of the Earth) – regarded as a more conservative publication than the *Essays* – it was recognised that the findings of science and history and their implications for theology and the Bible must be accepted.

Evolution was coming to be regarded as the method of creation. Had not Darwin said that 'by nature' he meant the laws ordained by God to govern the universe? Would it not reflect greater glory to God if he had designed all things to make themselves? The Revd Charles Kingsley (1819–75) impressed Darwin with his argument that God could have created animal forms capable of self-development rather than divine intervention to create distinct species and, with Kingsley's permission, Darwin included it in the preface to the second edition of *The Origin of Species*. The American biologist Asa Gray (1810–88) argued that design should be applied to the evolutionary process in general with the Creator using it to work out his plan.

Darwin looks at the world and exercises a value judgement: God cannot be moral. Paley's God of divine wisdom and goodness was under sustained attack. The latter impresses Darwin when he looks at the beauty of the world but the former when he thinks of the dark side of nature. Suffering does not always improve us and there is much that is pointless. If God is good and all powerful why does he allow it? But in exercising a value judgement Darwin has forgotten that you cannot base an 'ought' on an 'is'. He has described a process and now he says it ought not to be like that. Why? Because of the values he has inherited from Christianity or humanism. In both, cooperation and altruism are praised but these are values, not objective facts, and others stressing them may still postulate a good designer. J. S. Mill said that nature commits crimes every day for which men are hanged or imprisoned so it cannot guide our actions. Vice is more natural to us than virtue, the good qualities are a victory, that is, selfishness is natural but self-control has to be learned. However, the problem of evil is a strong argument against a moral God and it had been given additional force by the evolutionary theory. We will return to this problem in later chapters.

Nineteenth century scientists were so inspired by the progress of science that Hubert Spencer saw humanity moving upwards morally and letting 'the ape and tiger' die. He painted an optimistic picture of the future as the 'hangover' of evil disappeared. Wars would soon be eliminated and the brotherhood of man realised. Such optimism turned to ashes in the 20th century wars where 'civilised' nations committed barbarity comparable or even worse than the dark Ages.

However that may be, it is clear that arguing for the presence of God in the world after Darwin was more difficult than before, for his work confirmed what we saw in the last chapter that God might be acceptable as a remote someone 'outside' nature but not interfering with its workings. The 'absentee landlord' could maintain transcendence but not immanence. Initial and natural causes had replaced final ends. The world of the middle ages with its mystery and mysticism and divine intervention had increasingly given way to the natural and abstract world of science. What Darwin had done was to reject the analogy of the watchmaker and the watch which is external design imposed from without. The theist, as we mentioned, could still argue that the Creator had built design into the 'very structure of the process through which higher forms of life and eventually man could come into being'.[37] And it might also be maintained, that it had been assumed that man's descent determines his nature, that is, that source fixes meaning which is genetic reductionism. It is materialism finding significance in primitive beginnings which is a philosphical assumption destructive to the dignity of man.[38]

Progress and evolution were the key concepts of the latter half of the century and the new historical criticism paralleled that approach. If scientists had moved away from a static view of nature why should the theologians not reject a static, once-for-all revelation? If evolution was the key to understanding the development of life forms why should we not see an

evolving image of God portrayed in the scriptures? The liberal
theologians began to understand inspiration not as applying
to the words of the Bible but to the writers. God's revelation il-
luminates the minds of his servants so that they think out the
truth for themselves. Revelation is progressive and depends on
the ability to receive it and when it is written it clothes itself in
the thought-forms of the time. Thus the errors of scripture are
not to be ascribed to God but human fallibility which is seen
in ascribing the killing of the Amalekites to the command of
God (1 Sam.15:3) or the execution of witches (Ex.22.18). This
is one answer to Darwin's God as a revengeful tyrant. Such a
picture of God belongs to the lower levels of understanding of
what God is like and is ruled out by the later developed revelation of God.[39] We have evolved and so do our ideas about God.

Religious experience was stressed with literary and historical
criticism analysing it. The approach followed the scientific
method of induction. In the old view the theologian deduced
necessary conclusions from the propositions written in scripture, arranged them in a system, and arrived at revelation. The
new approach considered the evidence of man's religious consciousness and inferred a Mind behind the books of the Bible.
How could we explain the astonishing unity of the Bible which
contains books written over such a long period of time and by
different authors without a central Intelligence?[40]

With regard to science its prestige and confidence had
reached new heights and it promised a future of hope for humanity. The 'nobler race' predicted did not emerge but rather
the survival of the fittest. This might have been predicted,
since its basis was the mechanistic and naturalistic world view
which persists today.

3 After Darwin

Faith (in God) means blind trust, in the absence of evidence, even in the teeth of evidence … the theory of evolution is about as much open to doubt as the theory that the earth goes round the sun …

Richard Dawkins

The quotation above reflects the confidence of some scientists today in the theory of evolution with natural selection replacing a creator. But Richard Dawkins is irritated that, unlike the fact that the earth moves round the sun, he has to defend evolution which has been caricatured on the ground that chance and random variations could not have produced complex beings. But, assuming that the theory is correct, we want to know if it rules out the presence of God in the world. To see if this is the case we will examine various theories of the origin of life, discuss inheritance and genes, consider Dawkins' views, and show that there is dispute about how evolution has occurred. We also continue the debate about a Designer and the involvement of God in the process. Concerning what we are, the old question surfaces: is it nature or nurture that makes us: genetic inheritance or the information transmitted by the culture of our society? Can we control our genes or are we determined by them? What relation have the genes to the culture which we have imbibed? Currently these issues are being debated and have practical implications as we move into the new century.

HOW DID LIFE BEGIN?

The neo-Darwinian belief is that the earth is 4600 million years old and life emerged spontaneously in the sea about 3000 million years ago. All around us there are many species but they are descended from one or a few organisms which evolved in the sea. The method of evolution is random genetic mutation and natural selection. Probably life began in the sea through the combination of amino acids and proteins as the chemicals experienced electrical discharges from lightning and the impact of ultraviolet light. Attempts have been made in the laboratory to induce the same conditions as this 'primaeval soup' but without much success. It appears that nobody knows exactly how or where life began. Some think here, while others believe that the 'seeds' of life came from outer space. As to how it happened, scientists refer to the actions of blind physics and chemistry but there is disagreement whether the earliest life forms evolved from organic chemicals or inorganic minerals.

If life did begin in the sea there is an interesting parallel with the narrative of Genesis Chapter 1. It records the Spirit of God (ruach) hovering like a mother-bird over the primordial waters, the fons et origo, from which life emerges. The waters represent the formless chaos with the Spirit bringing order and ruach is the life-giving element of God. In the Bible the presence of God in the world is often symbolised by water: the Flood, the crossing of the Red Sea and the Jordan, the wedding at Cana and so on.

Fred Hoyle and Francis Crick have put forward an extraterrestrial origin. Crick gives us the fascinating picture of an unmanned rocket carrying primitive spores which is sent to the earth by a higher civilisation developed billions of years ago. Before we write this off as science fiction we need to remember that here is one of the men whose imaginative genius gave us the DNA helix where other biologists failed. Crick is conscious that the beginning of life was 'almost a miracle, so many are the conditions which would have had to be satisfied to get it going ... we can form no clear idea of whether such a soup is likely to lead to a primitive living system within a reasonable time – say a billion years – or whether most of these soups are doomed to remain lifeless almost indefinitely because the origin of life is such an exceedingly rare event'; we must approach the origin of life as 'one of the great mysteries'.[1]

But there are objections to Crick's proposal. It transfers the problem of the origin of life elsewhere, the theory looks like science fiction, and where are these beings who could design such a rocket? If they are there why have we not heard from them? With such advanced knowledge we would expect them to have spread all over the galaxy. Crick has no convincing answer to the last objection. He admits that it would have been a demanding undertaking to construct spaceships and set up colonies. His other reason for these aliens not colonising is that they may have grown tired of technology and adopted a different life style perhaps even cultivating a purely spiritual way of life. What has happened since Crick's speculations is the discovery of a solar system similar to our own which could contain extraterrestrial life. It is in the constellation of Pegasus and its sun, a star called 51Peg is 40 light years away or 240,000 billion miles – a short distance in astronomical terms. The question is: is there life there? The new solar system is based around a star which is like our own sun and that increases the chance of life existing. The star can be observed from the earth with the naked eye, but if we can observe them they can see us, so why have they not made contact? Their solar system is five billion years older giving them a considerable start on us. It raises again the question that baffled Crick.[2]

Andrew Scott thinks that elementary textbooks of biology relate a simplistic tale about the origin of life on earth which may be true, or partly true, but may be quite false. As a biologist, he believes it is too simple to be taken seriously: 'I am not able to reveal to you how life began, because I do not know for sure, and neither does anyone else.'[3] Paul Davies, a physicist, agrees: 'The origin of life, the evolution of increasing biological complexity, and the development of the embryo from a single egg cell, all seem miraculous at first sight, and all remain largely unexplained.'[4] Jacques Monod admits that the chances that the macromolecules would combine in the right way to produce life could be compared to a lottery, for 'the universe was not pregnant with life nor the biosphere with man. Our number came up in the Monte Carlo game.'[5]

Richard Dawkins, in *The Selfish Gene*, chose the theory that life began in an organic 'primaeval soup' but in *The Blind Watchmaker* he turns to the 'inorganic mineral' theory as the originator of DNA (deoxyribonucleic acid), which we explain below. The original life on earth could have been based on self-replicating inorganic crystals such as silicates followed by organic replicators and finally DNA. Self-replication is the answer but Dawkins admits that it seems improbable to him that 'randomly jostling atoms join together into a self-replicating molecule.'[6] He confirms that there is no definitely accepted account of the origin of life. One has to proceed on the assumption that the laws which operate just popped into existence like the universe itself, but such laws reflect order and can we have order without a designer? Hoyle examined the factors in the origin of enzymes which are essential to the functioning of life and calculated that the chances against their having originated by natural, random movements of molecules were immense. The odds are far higher than winning the national lottery many times. He concludes that there must be a Mind behind the universe.

GENES AND DNA

Are we determined by our genes? James Watson says: 'We used to think our fate was in the stars. Now we know, in large measure, our fate is in our genes.'[7] Intelligence, sexuality, illnesses, talents and many defects are inherited through the genes, hence the current focus on gene research and engineering. In America, there has been the founding of a multi-billion-dollar Human Genome Initiative project to map every gene of a prototypical human being on the DNA. Some, while recognising the project's value with regard to disorders, point out the dangers, stating that many of the disabled people who today live quite happy lives might never have been born since it was predicted otherwise on the basis of their defective genes. The project could also threaten privacy and civil liberties and lead to discrimination in many areas.[8]

What is a gene? Darwin had problems about heredity and somewhat reluctantly accepted Lamarck's view of acquired characteristics in the sixth edition of *The Origin of Species*. He was unaware that someone existed who had done work on heredity: Gregor Mendel (1822–84). It is ironic, considering the difficulties that Darwin had with religion, how much help he might have derived from a monk, but Mendel's papers were not discovered until 1900. He had experimented in his monastery, crossing tall and dwarf peas, and discovered that in two generations the tall peas were about three times as numerous as the dwarf. Further crosses showed the real nature of the genotype (genetic structure) in contrast with the phenotype (appearance). He was looking for the laws governing the inheritance of simple characteristics in plants, for example, wrinkled or smooth skins of peas. Having established ratios through the generations, he concluded that there were units of inheritance that control the process of transmission. These may be dominant or recessive and in humans result in our characteristics. Every living organism is made of cells and when they divide into chromosomes transmit heredity.

T. H. Morgan (1866–1945) continued the investigation of the transmission of Mendelian factors and pointed out that they were like beads on a string of chromosomes: he called them genes. Variations may be caused when an inexact copy of a gene, that is, a mutation, occurs. It could be helpful or harmful, but if the former, novel genetic potentialities appear. Then the study of genetics took a leap forward in our century when the actual structure of the chemical carrier of the genetic code, deoxyribonucleic acid or DNA was discovered by James Watson and Francis Crick.

DNA is shaped like a double helix and consists of long chains of molecules in specific sequences which determine characteristics and behaviour. The basic units of the chains are called 'bases': A for adenine, G for guanine, T for thymidine and C for cytosine. These letters are the genetic alphabet and every child gets a set of DNA from each parent. It is remarkable that the bases in DNA 'recognise' each other and have a physical affinity that results in their tending to stick together so that A goes with T and G with C. It happens because they have complementary shapes.[9] But mistakes can occur in the copying process as the DNA molecules are produced for the new offspring or by radiation impact or other accidental occurrences. The novel codes result in random variations and natural selection works on these. In order to combat the accusation of blind chance, it is contended that natural selection builds good design by rejecting variants and accepting those that improve adaptation to the environment.

The modern version of the theory of evolution occurred from 1930 onwards and stressed genetic inheritance. Genes confer an advantage or disadvantage as they are transmitted from parents to offspring. The beauty and structure of the DNA helix reflects design but it is thought that the novel codes produced are automatic and happen by chance. The DNA code explains variations, but mutations are in addition to the usual reshuffling and result in a totally new characteristic appearing in the offspring. There is a gene pool and the appearance of something not normally contained in these genes would be a complete novelty but it must interact with others. Neo-Darwinians contend for small mutations which occur from generation to generation and modify genes in a helpful manner rather than large mutations which can be harmful.

The cells are the chemical nature of genes and are composed of DNA or in some cases of RNA (ribonucleic acid). Mutations enable life to survive and produce varieties from a single species and varieties in due course become species. Without these mutations homo sapiens would not have evolved but it is difficult to accept that we are here partly because of mistakes in the DNA coding. Genes have to be organised so that they resemble computer files of information but the materials used are not disks, rather the DNA. Thus we have information carriers that work on the DNA material. It is fascinating to think that the information-bearing pattern endures through all the changes of our bodies and transmits to our offspring. Some have seen it as the soul which could be transmitted to a new body when we die.

Genes act within the cells so we cannot give them an isolated role and place too much stress on their determining what we are. The DNA chemical is present in all living organisms with most of the work being done by the proteins which are called amino acids and are responsible among other things for chemical changes. Each of us resembles a miniature factory! There are 23 pairs of chromosomes in our cells with the 23rd pair in two shapes. In the female the shapes are the same, hence she has XX chromosomes, but the male has XY chromosomes because the shape of one is much smaller and has fewer genes. Some females have a Y chromosome yet are born females because the Y does not function. The female gets one X from each of the parents but the male receives the X from the mother and the Y from the father.[10] Genes can now be transferred from one animal to another and it is stressed that there is little difference between our genes and that of the chimpanzee.

According to Richard Dawkins, the organising principles are the selfish genes which compete with one another for selection. The genes in his writing take on a form of life being intelligent, selfish, ruthless and competitive. They have powers to 'create form', to 'mould matter', to 'choose' and 'aspire to immortality'. Dawkins defines a gene as any portion of chromosome material that potentially lasts for enough generations to serve as a unit of natural selection.[11] It plays the role because of its copying ability, its fruitfulness and its longevity. A gene is good when it enables the organism to survive, but bad when it fails to fulfil the function. Ruthlessness is a virtue, altruism is not, the latter only operates for selfish ends. Dawkins does seem to be giving a purposive role to genes for their goal is replication, and in stressing their selfishness builds a case for the theological doctrine of original sin. It is based on a causal connection between the sin of the first human and the sin of all mankind (Ps.51:5; Rom.5:12).

At least that was the way some clerics interpreted the transmission of the selfish gene and Dawkins records their letters of appreciation! Selfishness is built-in but altruism, he says, is not in our genes and will have to be taught. Why? If we have a capacity for evil why not for goodness? Of course we must teach morality but some people are more sensitive to right and wrong than others. Yet they may not have experienced a good environment. How can it be explained without referring to their nature? Dawkins admits that genes cooperate to build bodies, and selection favours those that do so, but that may mean that these genes are sacrificed.

A change takes place when Dawkins reaches Chapter 11 of *The Selfish Gene* where he points out that the genes despite their importance are too narrow a basis for understanding evolution and proceeds to coin the term 'meme': the unit of cultural transmission. They help or limit possibilities, that is, defective genes, but are part of a process in which there are many parts. Dawkins' subsequent attention to cultural evolution sees it taking over from the genes and becoming the new replicator.

Dawkins' theory differs from the sociobiology of E. O. Wilson that asserts the dominance of genes over culture. Wilson insists that culture only offers the genes protection whereas Dawkins contends for some kind of independence with genetic interaction. The 'memes' or culture items contribute to gene-replication or changing gene frequencies.[12] Thus we have two views of evolution, and it is debated whether the observed characteristics of an organism (the phenotype) can tell us more about it than the genetic basis (the genotype). Cultural transmission is by information and learning and depends on the sophistication of a society. But what is acquired by a person cannot get back into the genes – no transmission of acquired characteristics – and be passed on, whereas the genetic is transmitted by reproduction.

The strong form of sociobiology teaches that genes transmit information and build bodies as protective coverings. What arises in culture safeguards the shields, and Wilson readily accepts that religion is one of these protectors since it has prohibitions regarding sex, the family, nurture of children, food and so on. Wilson thinks that although religion will survive for a long time, we need a better one which cannot be supplied by current religions because they are built on the myths which science has destroyed. We are reminded of what was said in the last chapter about a religion based on science. Wilson accepts that religion persists because the genes created the capacity for religious behaviour. It would seem that he is thinking of a religious gene. Does this show a distinction between the human and sub-human animal since it is only the former that engages in such activity? And if there are genes that have such religious tendencies then have we something innate? The difficulty with this argument is that we are told that the chimpanzee has 99 per cent of the same kind of genes as the human being, but since there is such a difference between the chimpanzee and the human the one per cent must be very special.

What is being said about genes means permanence or necessity. In the blood cells there are strands of DNA and the patterned molecules stretch back to the single micro-organism where the universal genetic code took its form. But change is brought about by various factors: mutations, the recombination of genes and genetic flow caused by migration. There is also genetic drift in a small population through failure of parents to transmit genetic instructions to offspring.[13] Wilson thinks the genes take from culture what enables them to survive, but John Bowker asks: why may the selection not include values which are true, beautiful and good? We have these values, how did we get them if they do not contribute to the benefit of a society? Do they not have a history independent of the genes? We need something to survive but also to enhance life and while these things are not identical they overlap. Dawkins does not tell us what gives survival value to his memes.[14]

Dawkins, however, has had a lot to say about religion, contending that the idea of God replicates itself by the memes: spoken and written word, music and art. It has survival value because of its psychological appeal, providing a superficial answer to questions about life and hope of another existence where injustices will be rectified. It is a dangerous meme: a virus which has got into the brain. But the argument can be countered by pointing out that, in origin, the ancient religious traditions both East and West, had no belief in a worthwhile life after death. And, does such a meme not produce much that has been good? Why concentrate on the violence associated with religions and ignore their saints?[15] Beliefs about how the world works have survival value but not all of them are based on this. There are values such as honour, justice and truth for which people have been prepared to die. Celibacy does not contribute to the survival of anything. There are values which are instrumental for the good of something else but also intrinsic ones which are adhered to for their own sake.

Dawkins defines faith as blind trust, in the absence of evidence, even in the teeth of evidence, and cites the story of Doubting Thomas (Jo.20.24–9). Thomas should be admired, according to Dawkins, since he demanded evidence unlike the other disciples. They exhibited the meme of blind faith which secures its perpetuation by the expedient of discouraging rational inquiry.[16] But Dawkins is not correct for, according to the record, the other disciples had evidence: Christ had appeared to them, and they were sceptical when they first heard the story of the appearances (Mk.16:12; Jo.20.10–23). Since Thomas had been with the group for a long time he should have believed them but he wanted to see and experience for himself. Faith in the religious sense is based on experience from which arises the interpretation. Science is also based on experience and A. N. Whitehead sees a parallel:

> The dogmas of religion are the attempts to formulate in precise terms the truths disclosed to the religious experience of mankind. In exactly the same way the dogmas of

physical science are the attempts to formulate in precise terms the truths discovered by the sense perceptions of mankind.[17]

We will discuss revelation later in the contex of the religious traditions.

DESIGN

Dawkins sees no need for a Designer and postulates the blind watchmaker: natural selection operating by means of chance and random variations. He thinks that we are here because of a long chain of accidents, guided only by the natural selection of favourable combinations. This has produced all the ascending variety and complexity of life from amoeba to man. The most complicated thing does not come into existence suddenly but by slow steps: cumulative selection. Chance could operate in small changes and cumulatively lead to large. Natural selection is the blind watchmaker, blind because it does not see ahead, does not plan consequences, and has no purpose in view.[18]

Evolution takes time to evolve complex things and while mutation is random natural selection is not. Dawkins does not like the use of the word 'chance' since it indicates something of a fluke. He prefers what he calls 'non-random survival', which seems to imply some necessity in the shuffling operations of chance. But necessity implies order. Polkinghorne criticises Dawkins for not going on and exploring why the small steps of evolution result in systems of such wonderful complexity as ourselves. It would not happen in 'any old world' but in one that had a divine purpose expressed in the given structure of the world.[19] In an evolutionary world things are flexible and fruitful so various possibilities could be realised as it develops itself.

We might think of the world as God's experiment with 'chance', shuffling or exploring possibilities and bringing into being novelties: mutations. Necessity would express the natural order of the world, that is, reliable transmission of genetic information from one generation to the next, and set limits to chance. It implies processes which trigger instabilities and random events, leading to a sequence of uncorrelated possibilities.[20] A. R. Peacocke sees the potentialities of the world being unveiled by chance and compares evolution to the improvised development of a great fugue. But Polkinghorne thinks it is creation itself which is doing the experimenting. To say, however, as Dawkins does that the process is blind is a metaphysical claim that should not be presented as if it were a scientific conclusion.[21]

Dawkins uses an example from the seaside. The smaller pebbles run along the length of it whereas the larger ones are in separated zones. This occurs not because there is some Designer but by the blind action of the waves. Similarly in cumulative selection reproduction there is a sorting over many generations in succession. In this sense it is non-random. Cumulative slow change is crucial to Dawkins' case as it was with Darwin.[22] A long time span is needed in which chance, luck and coincidence will enable cumulative selection to string together random mutations in a non-random sequence to produce mankind. But Dawkins admits that it is difficult to explain how the process got started. It cannot work unless there is some machinery of replication but the only one we know appears to have come into existence by many generations of cumulative selection! The argument is circular. Dawkins rejects a Designer because we cannot explain how he originated but presents us instead with a theory which depends on 'sheer naked coincidence, sheer unadulterated miraculous luck.'[23]

What the neo-Darwinian has to explain is how conditions arose so that life could originate, and whether probing into the lower forms of the development in organisms can explain the complexity, organisation and principles which govern the whole. Dawkins dismisses Lamarck's acquired characteristics, instantaneous creation, ideas about God in the process, and rests his case on cumulative natural selection. It is the ultimate explanation for our existence which, as he says, dissolves astronomical improbabilities and explains prodigies of apparent miracle. We are simply complex machines governed by biological and chemical laws. He and others want to extend their reductionism so that chemical laws can be derived from those of physics, biological laws from chemical, and the laws of psychology and sociology from biological. Psychology reduces our behaviour to instinct, conditioning or the traumas of childhood, and physiology reduces thoughts to electrical signals in nerve cells and emotions to glandular secretions in the blood stream. Dawkins admits, however, that teleology has been replaced by teleonomy: '... in effect teleonomy is teleology made respectable by Darwin, but generations of biologists have been schooled to avoid teleology as if it were an incorrect construction in Latin grammar, and many feel more comfortable with an euphemism.'[24]

Jacques Monod shares in this admission and points out that a sense of purpose is common to all living beings. The concept must not be rejected, as certain biologists have done, but recognised as essential. He sees the distinction between living beings and other structures or systems by the property of teleonomy.[25] But we might compare necessity, order and law, as seen in the operation of the genes, with the written description for building a house or ship. The genes are written in a code of a few symbols but the order can carry an infinite number of meanings. There is evidence of plan because of the difficulty of accepting that a disorganised collection of molecules could assemble themselves in the whole of a living organism. It is still a mystery to biologists how it can occur, for example, in the embryo in the womb developing from a single fertilised cell. In the process many cells have detailed to form liver and bone nerves at an astonishing level of accuracy. Even if damage occurs new cells can replace mutilated ones, or those that have been displaced can find their way back to the right position:

> It is difficult not to think of a blueprint or plan of assembly, carrying the instructions needed to achieve the finished form. In some as yet poorly understood way the growth of the organism is tightly constrained to conform to this plan. There is thus a strong element of teleology involved. It seems as if the growing organism is being directed towards its final state by some sort of global supervising agency.[26]

THE DEBATE: IS EVOLUTION SLOW OR DUE TO SUDDEN LEAPS FORWARD?

Dawkins' slow cumulative development has been challenged by Niles Eldredge and Stephen Jay Gould who favour the jump (saltus) theory. They are palaeontologists who see large-scale biological systems (organisms, populations, species, higher taxa) and the interrelationship among them as crucial and recognise the fixity of species for millions of years. Darwin said that species are not discrete entities but 'permanent varieties': stages in a continuous evolution. But he did see the gap between species, that mating does not take place across species boundaries, and in the sixth edition of *The Origin of Species* acknowledged that species tend to remain stable. However, we know that there are species present today which were not here 20 million years ago.

Eldredge thinks that culture is unique to mankind and Dawkins recognises it: learned behaviour transmitted through language rather than genes. But what needs stressing is that it makes our society different from the sub-human. Eldredge refers to one of E. O. Wilson's books, *Sociobiology*, which contended that behaviour, from schizophrenia to religion to criminality, was determined by genes. But there is little mention among such ultra-Darwinians that economics and reproduction are interrelated but have their different goals. What occurs in the development of life is that females promise sex and males provide the food, thus economics and reproduction are related. But homosexuality, for example, shows that all behaviour is not for reproduction and adoption indicates that we are not intent upon making copies of our genes. Life is more about staying alive than reproducing ourselves. But in any case making copies of genes is a postulate about the future and therefore a goal.

Eldredge emphasises holism rather than genes and complex interacting ecological processes. Bodies have parts: organs, tissues, cell, molecules, atoms. The atom itself is composed of elementary particles but it is a complex system. An ecosystem could be a forest or an open field and the significance of the whole is neglected by concentrating on the parts: unable to see the wood for the trees. Species are wholes. Nor can we think of an organism apart from societies or populations. As we have seen, natural selection works on hereditary factors: genes which dictate the synthesis of proteins. The interaction of these proteins in an environment where the effect of culture and physical conditions will result in a phenotype: appearance and behaviour. Such culture will affect genes. Any one individual represents a small sample of the factors present in a population and changes can take place in the relative frequency of genes within it. The naturalists think of natural selection as only a filter and lay stress on the genetics of populations where gene changes occur through chance factors: genetic flow and drift. The flow arises from the migration of people and the drift means that genetic instructions can be lost from a population through failure of parental transmission. It usually happens in small rather than large societies. The latter preserve characteristics of individuals since they are carried by so many but, for example, in small communities such as the American Indians, blood group B has been lost.[27]

Saltus is punctuated equilibria where stasis (no change) is interrupted by brief bursts of evolutionary change resulting in large-scale evolutionary patterns and species selection. The debate then is between macro- and micro-evolution, with the former meaning evolution on a large scale: origin of major groups such as mammals. It is based on the need to explain the abrupt appearance of some things, for example, bats and whales.[28] In reply Dawkins says that cumulative change is more probable than a tremendous jump: macro-mutation. He points out that small mutations occur when pieces of the genetic code get scrambled as the genes are being copied perhaps by one letter replacing another or a chunk of DNA being accidentally cut out and spliced in somewhere else or a piece of chromosome getting inverted. But he admits that there are mutations which cause large and perhaps sudden changes: snakes have many more vertebrae than their ancestors.

He also thinks that the palaeontologists are making a mistake based on the gaps in the fossil records which make the transition from ancestral species to descendant species appear abrupt and jerky. Dawkins believes that such gaps can be explained by recognising that evolution occurred in a different place from the location of most of the fossils. We are looking for the links in the wrong place! Dawkins chides Gould for giving the impression that he was opposed to neo-Darwinism in its 'gradualism' form. The media reported that Gould's work resembled the old belief in catastrophism which saw the fossil record as reflecting a series of discrete creations terminating in mass extinction: Noah's flood. All that Eldredge and Gould are saying, according to Dawkins, is that there are brief bursts of activity rather than a constant rate.

How then does he explain Gould's statement that neo-Darwinism is effectively dead? Dawkins says that it provides no basis for any lapse in morale and refers to his belief in a higher form of selection which they call 'species selection'. But a consideration of the writings of Gould does not support what Dawkins is saying about him. Concerning Dawkins' book, *The Selfish Gene,* Gould disgrees with him, arguing that he is still making genes primary and that bodies are only there for the sake of the genes. Dawkins thinks of bodies as machines used by genes and tossed away on the scrap heap once genes have replicated. But genes cannot be seen, they are hidden away in DNA, hence natural selection works on what can be observed, namely bodies. It favours some because they are bigger and stronger or more attractive and rejects others.

Gould like Eldredge is saying that selection acts on wholes not on parts: selecting wholes or organisms which because of the way their parts interact confer an advantage on them. Bodies cannot be atomised into parts each constructed by an individual gene. Dawkins, according to Gould, is a reduction-ist who breaks things down into parts and thinks that he un-derstands wholes by this method. It works for simple objects with few parts but organisms are more than a collection of genes. Gould writes: 'They have a history that matters; their parts interact in complex ways. Organisms are built by genes acting in concert, influenced by environments, trans-lated into parts that selection sees and parts invisible to selection.'[29]

The fossil record has abrupt transitions and we cannot imagine the intermediate forms which gradualism requires. What use is half a jaw or eye? Hence the palaeontologists contend for discontinuous evolution in that major changes can occur through genetic alterations and rapidly spread through a population.[30] But Darwin equated the slow with the natural and the fast with the supernatural, and Dawkins follows his master for he does not want to make that admis-sion. But slow cumulation is just as difficult an explanation as sudden, for much could go wrong in the various stages.

ALTRUISM

Sociobiology sees altruism as stemming from kinship. It is a
form of selfishness as each group looks after its own and its
territory, on the basis of all for one and one for all. A bird
may warn others of a predator and in so doing lose its life but
it means that the kinship group can survive and their genes
be transmitted. Altruism has the selfish purpose of preserving the group. Dawkins sees the benefit of altruism and cooperation and the need to defy the selfish genes of our nature
and the selfish memes of culture. But it does not exist in
nature: 'We are built as gene machines and cultured as meme
machines, but we have the power to turn against our creators.
We, alone on earth, can rebel against the tyranny of the
selfish replicators.'[31] Thus he modifies his gene determinism
and says that they only control us in a statistical sense and
there is no reason why other influences cannot override
them. And he is not advocating a morality based on evolution.
It is the strong form of sociobiology that gives such primacy
and continuity to the genes in determining both culture and
what we are. But units of selection are much wider than that
as we have noted: gene-complexes, organisms, populations,
genomes and species.

   J. S. Haldane (1860–1936) jested about the narrow kinship
view of altruism: 'Greater love hath no man than this, that he
lay down his life for two brothers, four half brothers or eight
first cousins'.[32] But of course there is more. There is sacrifice
for country, good causes, companions and friends, and if the
Christian doctrine of the atonement is correct for one's
enemies. We have to go beyond sub-human nature if we are to
understand how altruism functions in various societies which
means that a one-sided stress on genetic transmission by reproduction needs supplementing by cultural transmission
which will enshrine what religion, philosophy, psychology, sociology, can teach us about morality. If morality is concern for
others then altruism will have a prominent place. Dawkins
admits that genes cooperate in building bodies but in order to
replicate and assist mutation, hence he finds it difficult to
dismiss altruism even at this level. The religions recognise altruistic tendencies but if selfishness is built in, insist that our
nature needs changing by the grace of God.

IS THERE A CREATIVE PRINCIPLE?

Theists have thought of creative forces in evolution such as a nisus or urge or organiser but this vitalism or neo-vitalism has not impressed people like Dawkins, Vitalism is in the tradition of Aristotle whom we recall argued that forms were inherent in objects and caused them to take their characteristic shape. Forms do not change but bodies do. God was the source of the forms and these could be equated with souls hence Aquinas made the connection with the biblical understanding of the soul. It is the soul which is the source of the purposive activity of an organism. In order to escape the charge of vitalism theologians have seen God as the primary cause of all things but working through the secondary, that is, the natural, so that the operation is internal not external. Those who contend for 'leaps' in the progress do not postulate a creative principle but the debate shows that the question has arisen again and the danger is recognised by Dawkins. He is dogmatic in his denial but not all biologists agree with him. They go as far as to give the same kind of mechanistic explanation of life, saying that it is simply the overall result of many thousands of chemical reactions proceeding at appropriate times, in appropriate places, at appropriate rates, but do not rule out the possibility of such a principle. Andrew Scott writes:

> It remains possible that there may be some mysterious principle behind the origin of consciousness and thought and our apparent free will, and that principle might either be physical or spiritual in origin; but, on the other hand, there might be no such hidden principle – most scientists do not seem to believe there is.[33]

But biology does not speak with a single or oracular voice. There are those such as Richard Lewontin who point to the importance of consciousness as giving rise to history, personhood and society. Consciousness is an aspect of the structure of organisms, not simply genes. With regard to design, Stuart Kauffman speaks of a mysterious propensity towards order, large scale and small scale, in the living and non-living universe as well as in mathematics.[34] The biological explanation does seem incomplete and we need to draw on other disciplines to try and understand the complexity of the human being.

HIGHER AND LOWER LEVELS

There is another way of seeing God acting in the world which arises from the discussion of wholes. In an organism there is the emergence of new levels so that we can speak of the higher and lower. With each new level properties appear which, though related to the lower, cannot be explained by it. They are irreducible properties that arise out of the organisational complexity: physiologists, for example, can examine the parts of the circulatory system but they cannot explain the whole in terms of these. A living organism functions because a coherent pattern is imposed on the parts of which it consists: the organs are subservient to the organism, the tissues serve the organs, the cells serve the tissues, proteins and other substance serve the cells. But it is the higher that controls the lower levels, for example, when the cells around a wound stop growing when it is healed. If the organisation cannot control the cells which have become diseased the result may be death.

How is this ordered functioning in which the whole controls and is more important than the parts to be explained? We cannot just speak of entities as the sum of their parts for the whole influences the parts in a sort of 'downward causation'. How it occurs still eludes the scientist but without it there can be no complete explanation of embryology, evolution or the origin of life itself. But it is not possible in advance from any analysis of the parts to predict the kind of emergence that is going to occur. There is a fundamental unpredictability in nature, as we will see in Chapter 5. Complex wholes and emergent properties are part of the scenario. The fact that there is some pattern or principle or ordered function in organisms which, while related to the parts appears independent, has made some theologians think of it as analogous with God in the world and yet beyond it: panentheism. The concept of 'downward causation' is also fruitful, not in the sense of God disrupting natural cause and effect but as we act in connection with our bodies. We are conscious of being a self which expresses itself through the body but has a sense of transcendence over it and we might think of God's transcendence over the world in a similar way and expressing his intentions within the causal nexus.[35]

CONCLUSION

The work of Gould and Eldredge shows that while natural selection is important it is not the only factor in evolution. But natural selection replaced God in the process and neither Darwin nor Dawkins like this 'punctuated equilibrium'. We are wary of seeing God as being behind such fast saltatory genetic changes but would make the point that it is unwise to try and explain something by one major factor as Marx tried to do, that is, economics. Other factors, such as divine influence, might operate alongside natural selection and the saltatory. The process has resulted in a creature who has values and these could not have been predicted from our origins. Genes do not produce character or will. It is in the study of the inner life with its thoughts, feelings and willing, and in the external interaction with others, that we are likely to understand values and what gives meaning to life. We aim at ideals so purpose is connected with the future. The pull of the future is recognised by philosophers of science. Karl Popper points out that Heisenberg's theory of indeterminacy (discussed in Chapter 5) has made the clockwork picture of the universe imprecise. We have the mathematical probability theory which implies a propensity inherent in things enabling them to realise a goal. New possibilities or emergents arise which are not predictable on the basis of past cause and effect. Popper says: 'It is not the kicks from the back, from the past, that impel us, but the attraction, the lure of the future and its attractive possibilities that entice us; this is what keeps life – and indeed, the world – unfolding.'[36] He does not rule out chance but couples it with the 'preferences' of organisms and their ability to achieve possibilities.

We conclude that neo-Darwinism has not been able to exclude God from the world and we think it will become more evident as we proceed. To develop this we move in the next chapter from biology to physics to see how the world view has changed from the picture presented by Newton.

4 Einstein

The most incomprehensible thing about the universe is that it is comprehensible … in every true searcher of nature there is a kind of religious reverence … science without religion is lame, religion without science is blind.

Albert Einstein

Albert Einstein was the greatest scientist of the 20th century but like Darwin his early years were not distinguished. When he was a child he spoke hesitantly and did not like the regimental style of German schooling. After his family moved from Germany he attended a school in Milan, Italy, but his interruption of lessons with questions and arguments led to his being asked to leave. He left without an entrance diploma to a university and subsequently failed the entrance examination to Zurich's Polytechnic. But he did better at the school in Aarau and was admitted to the Polytechnic where he continued to show his dislike of the formal instruction. It was no surprise that on graduation, the professors refused to recommend him and he had to take a job at the Swiss Patent Office.[1] He was an unassuming genius who paid little attention to the etiquette of society or dressing well. There are many stories about him in this connection. When he was living and working at Princeton he called at a friend's house and his wife answered the door. She did not recognise Einstein and looking at his shabby clothes, assumed that he was a tramp: 'Sorry,' she said, 'we cannot give you anything today', and closed the door. On another occasion when his wife, Elsa, expecting visitors to arrive, tried to get him to wear a suit, he said: 'When they arrive you can open the wardrobe and show them my suit'!

In this chapter we want to look briefly at the new view of the world resulting from Einstein's work, our place in it, and how God might be related to it.

TIME AND SPACE

Einstein changed the scientific perspective of time, space, matter, gravity and light. We think in a common sense way about time and space; we are aware of time passing and space is something we move around in and that appears immovable. Time means duration and neither it nor space has a relation to anything external. The understanding is based on Newton: 'Absolute, true and mathematical time, of itself and from its own nature, flows equably without relation to anything external. ... Absolute space, in its own nature, without relation to anything external, remains similar and immovable ...'[2]

These absolutes had been accepted for more than 200 years but Einstein challenged them being indebted to the work of Michael Faraday (1791–1867) and James Clerk Maxwell (1837–79). Hans Christian Oersted (1777–1851) in 1820 discovered a connection between electric and magnetic forces by passing a current through a wire which deflected a compass from the magnetic north. This enabled telegraphy to develop since the electric current could be used to deflect a magnetised needle somewhere else and pass on messages. The reverse was then tried and Faraday demonstrated that you could get an electric current from magnetism. He was an experimental physicist and became assistant to Sir Humphry Davy, head of the Royal Institution in London. Faraday, the son of a blacksmith, became the victim of the class system of the time: Davy's wife refused to eat at the same table with him and demanded that her husband do the same![3] This nonsense did not deter Faraday's work on electric generators and imagining what would happen when a magnet and current interacted. His proposal was that a magnet or a current-carrying wire sends out lines of force in a definite pattern depending on the shape and strength of the magnet or current. Hence it moved away from forces acting at a distance, as in gravitation. The space between the bodies was seen as the active carrier of the force.[4]

Maxwell converted Faraday's experimental work into mathematical form and showed that electric and magnetic forces moved through empty space at the speed of light. He predicted that these fields (lines of force, magnetic influence) under certain conditions would move like waves through space. Nowadays they carry radio and TV programmes and light is recognised as a form of the electric force, that is, electromagnetic. But sound requires molecules of air before it can be heard so electromagnetic waves would require a medium. Maxwell postulated ether as the medium and fields as modifications of it. But when Michelson and Morely in 1887 tried to measure the ether wind as the earth moved through absolute space they found that it had no visible effect. Einstein proceeded to dispense with the concept and accounted for electromagnetic phenomena by fields alone. A field is a region of space having physical conditions through which forces are transmitted, like ripples through an invisible jelly. Waves of light were propagated through the field in straight lines.[5]

The contradiction between Newton and Maxwell spurred Einstein into action. The field laws of Maxwell and Faraday showed that light was propagated at a constant unsurpassable speed but Newtonian mechanics taught that it was possible to increase the speed of an object indefinitely by adding more energy to it. Einstein agreed with Maxwell that nothing could travel faster than light and it was independent of the bodies emitting or receiving it. But at this point he realised that a principle of relativity established by Galileo might help him. This stated that all steady motion is relative and cannot be detected without reference to an outside point.[6] When we are travelling by plane there may be some turbulence but the motion is usually smooth and we do not feel the movement. But a glance out of the window shows that we are moving: an outside point of reference is required. However, nothing is at rest in the universe for the earth is moving relative to the sun and the sun is moving in space. There is no rest point for someone who is stationary because the earth is moving about 18.6 miles a second round the sun. What we have to do is to measure one thing against another: there is no absolute point of rest.

Speed is a measure of distance, a unit of time. But if the speed of light does not change then time must and that challenges the Newtonian absolute. Time would be relative dependent on whether we were moving or stationary. Light is constant, travelling at 186,000 miles (300,000 kilometres) per second and spreads through space with the velocity c, being independent of the emitting or receiving body.[7] But its finite speed requires time to get from its source to an observer hence there are no instantaneous interactions in nature. Common sense views of time and space are being upset. We are aware that time passes and we move around in space. It seems immovable and without content until we put something there. Space, we recall, for Newton was a kind of container and matter moved within it. We think of three dimensions: up–down, left–right, backwards–forwards, which evolve in time. But Einstein spoke of a four dimensional spacetime which does not evolve in time because time is not something which is separate from it. It is static and contains all of time, past, present and future. There is no absolute like the present which would divide the past from the future. Einstein was greeted with laughter when he first mentioned such a novelty and he did admit that thinking about it nearly drove him crazy! Ironically he once said:

> If relativity is proved right the Germans will call me a German, the Swiss will call me a Swiss citizen, and the French will call me a great scientist. If relativity is proved wrong the French will call me a Swiss, the Swiss will call me a German, and the Germans will call me a Jew.![8]

How could the distinction between past, present and future be an illusion? Perhaps a solution is that though all time is stretched out in a fourth dimension we experience it moment by moment in that we act out each instant. This is our common experience suitable for life and society but not for nature as a whole.

LIGHT, SPACE, TIME

Someone told Einstein that he should have been a clock maker since he was so interested in time. In 1887 it was shown that the speed of light was always the same no matter what the speed of an object. Let us think of a train travelling along a track with an observer on board at a central position who sends out a beam of light both forwards and backwards simultaneously. Another observer watches from the track. The train is the moving frame of reference and the railway track the stationary. The front door and the back door of the train can be

opened by the beam of light. When the beam is emitted the
person on the train thinks that both doors open simultaneously but the person on the track sees the back door opening
before the front door. The reason is that the person on the
track sees the back door moving to meet the light beam while
the front door moves away from the light pulse. Yet the speed
of light is the same for both frames of references. Hence each
observer has his own frame of reference and there is no
Newtonian universal time where events take place simultaneously. It is the relativity of simultaneity: signal, time, event and
observer exist in a relationship.[9] But where now is our past,
present and future? With Newton the separation between past
and future was objective in that it was determined by a single
instant of universal time namely the present:

> This is no longer true in relativistic physics ... There can no
> longer be any objective and essential (that is not arbitrary)
> division of spacetime between events which have already
> occurred and events which have not yet occurred.[10]

Speed is distance divided by time and if the speed of light is
always the same a problem arises about distance and time. An
example is given by Russell Stannard of an astronaut travelling
from earth to a distant planet. How long did it take him? The
astronaut and the mission controller at Houston, Texas, do
not agree on the time taken. If the spacecraft travelled close to
the speed of light, the mission controller would discover that
the astronaut's clock lags behind his own. At 9/10th the speed
of light the journey time as recorded on the astronaut's clock
would be approximately half that recorded by the controller.
Time then is not something which is universal and exists on its
own separate from space: 'future and past are just directions,
like up and down, left and right, forward and back, in something called spacetime'.[11]
    Of course we are talking about travel at high speeds where
there is a slowing down of the astronaut's breathing, pulse rate,
thinking processes and ageing. Thus the twin finds that when
he returns from space he is much younger than his brother who
has remained on earth. But we have not discovered the secret of
eternal youth for our mental processes have slowed down and
we will not be aware that we are living longer.[12] The fact that

moving clocks run slower is not connected with their mechanical properties but that space and time are different in moving frames of reference from those in the stationary. Matter will also be affected, a rod would shrink. Subjectivism seems to have entered physics since observers have different views of what is happening but the objectivity is the constancy of light.

GRAVITY AND LIGHT

Einstein challenged Newton's laws of gravity. We are conscious of inertia when the train at the station jerks forwards for our state of rest has been disturbed. The train remains stationary unless a force moves it and the greater its mass the more accelerating power is needed. We know the difference between throwing a cannon ball and a tennis ball. It is air resistance that stops cannon balls and feathers from falling at similar speeds but if a perfect vacuum was realised they would drop at the same rate. Newton thought that gravity works in proportion to the mass concerned, so the small mass objects had a little pull, and the large mass a strong one, and they would fall at the accelerating speed of 32 feet per second per second. Gravity counterbalanced inertia and its effect was transmitted through space instantaneously. But Einstein did not like this balancing act of gravity and the effect of inertia.

Newton thought that gravity was a force acting instantaneously at any distance between bodies but Einstein understood it as a magnetic field which acts on the bodies within it. Einstein imagined a man falling from the roof of a house and not feeling his own weight. Or if someone was in a lift when the cable snapped she would float weightless as the lift fell through the earth's gravitational field. An astronaut in a space ship with no gravity to hold his feet to the floor floats weightless but as the rocket accelerates its floor rises and makes contact with the astronaut so he would think it was gravity holding his feet to the floor. Hence Einstein became sure that gravity equals acceleration.

Suppose a ray of light crosses a lift from one side to the other while the lift is being accelerated. The far side of the lift moves upwards before the ray of light reaches it, which would mean that its wall would be hit by the light ray nearer the floor than the point at which it set out. The person in the lift would have seen a bend in a horizontal light which was due to acceleration producing the same effect as gravity when the lift was at rest. Newton thought that light was a stream of particles but Maxwell opted for waves. In 1905 Einstein showed that light consisted not of waves but of a stream of light quanta like minute machine-gun bullets – photons – so the particle view returned. But later it was shown that it could be both a particle and a wave. Light would be affected by gravity like anything else. Light travels in a straight path so that we measure its speed from A to B but if light were affected by gravity, time and space would, when viewed from within the gravitational field, be different. Without gravity the shortest distance between A and B which a light ray travels is a straight line but with gravity the line is not straight. Both acceleration and gravity can bend light so we have the principle of equivalence and can conclude that gravity does more than make things fall.

Einstein resolved the problem of the equality by replacing the gravity force by the curvature of spacetime and proceeded to try and understand the effect of gravity upon electromagnetic processes, the bending of the paths of light rays travelling near to massive objects. He went beyond Newton working out the interaction of gravity and electromagnetic signals and the relation between gravity and time. Matter creates a gravitational field and light is curved in it. In sum, gravity for Newton was a force operating instantaneously over distance, but for Einstein it was a function of matter and its effects transmitted in spacetime.

Gravity, matter, space and time are interrelated but if light is deflected by a gravitational field it is necessary to employ Riemannian geometry rather than that of Euclid in order to understand it.[13] A useful analogy in understanding the relation of matter, space and gravity is to think of space as a piece of rubber which stretches and expands and is affected by the matter in it. If the rubber is spread out on a table and a ball is placed in it and held up by the ends it becomes three dimensional. On the basis of the analogy, matter affects space, causing it to curve and gravity itself results from the curvature of space. Gravity is not just a force operating in a fixed background of spacetime but a distortion of spacetime caused by the mass and energy in it. We can put a number of balls in the rubber sheet and see how they distort it. The balls represent the planets and sun and they distort space according to the strength of their gravity with the sun having a greater effect on space than the earth because it has a stronger gravitational pull. Space is not only curved by the presence of matter, it continually expands, behaving rather like a rubber sheet that being pulled becomes larger and stretches so that objects placed in it move away from one another. Hence the galaxies of the universe are separating because of the expanding space between.[14]

We have moved a long way from the Newtonian fixed framework of space and time that was not affected by what happened in it. Bodies moved, forces attracted and repelled but time and space continued and remained unaffected. Now it was seen that spacetime and matter interacted as dynamic qualities. Space has an elastic quality since matter and energy can curve and distort it. It is like putting a heavy ball on a cushion which causes a curve and makes other balls roll towards it. The sun curves the space around it and other planets experience the curvature and 'roll' in the curved space. Hence gravitational force is replaced by the curvature of spacetime. Thus time is affected:

> … no universal, 'now', there is only 'here and now' for each observer so that space and time are joined together and are aspects of a single reality. And the structure of space cannot be considered apart from the matter embedded in it.[15]

General relativity explains gravity as a distortion in the geometry of spacetime. The planet Mercury is disturbed in its motion and Newton's theory of gravity could not account for this. But Einstein worked out corrections in the Newtonian theory as a result of the spacetime curvature and in 1919 during a solar eclipse it was demonstrated that light rays passing the sun's massive body were 'bent', thus confirming Einstein's theory of relativity. Sir Arthur Eddington put this historic moment into verse with a parody of the Rubaiyat of Omar Khayyam:

> The Clock no question makes of Fasts or Slows,
> But steadily and with a constant Rate it goes.

And Lo! the clouds are parting and the Sun
A crescent glimmering on the screen – It shows! – It shows!!

Five minutes, not a moment left to waste,
Five minutes, for the picture to be traced –
The Stars are shining and coronal light
Streams from the Orb of Darkness – Oh make haste!

For in and out, above, about, below
'Tis nothing but a magic Shadow show
Played in a Box, whose Candle is the Sun
Round which we phantom figures come and go

Oh leave the Wise our measures to collate.
One thing at least is certain, LIGHT has WEIGHT
One thing is certain, and the rest debate –
Light-rays, when near the Sun, DO NOT GO STRAIGHT.[16]

MATTER

Matter was affected, as we mentioned, by the speed of light. If it was possible to drive at 90 per cent of the speed of light, an observer from the stationary perspective would see the length of the vehicle decreasing but Newton thought that motion would have no effect on it. Matter, however, dissolves into energy while the total mass remains the same. Thus the famous equation: $e=mc^2$ where mass is equivalent to an energy e if c is the velocity of light. Mass was locked up energy and the shining of the sun is an example of mass-to-energy conversion. Atomic bombs and nuclear power stations demonstrate the equivalence of mass and energy hence we can no longer think of things as solid objects or separate from the forces that move them. It was this knowledge that led to the splitting of the atom and the bombs that destroyed Hiroshima and Nagasaki.

THE CHANGED VIEW OF THE UNIVERSE

The new view affects how we think the universe began: the 'Big Bang'. It marks the point when space and time were created and matter was small, hot and dense. About 15 million years ago a singularity occurred, density and temperature became infinite, and matter started to expand. Such expansion has continued so that we live in a universe which has galaxies moving away from each other at speeds proportional to the distance between them. If matter multiplied and gravity increased its strength the universe would contract in upon itself: the 'Big Crunch'.

Einstein apparently believed in a static universe though his work leaves room for flux and change. At first he refused to accept that the universe was expanding and introduced a controlling cosmological constant which counteracted at great distances the gravitational force. Later he admitted that it was his greatest blunder.[17] He had to accept the expansion because Alexander Friedmann and Edwin Hubble showed that the universe was changing with time in that distant galaxies were moving away from us. Matter could so curve a region in on itself that it became cut off from the rest of the universe: black holes had been discovered. Only if someone could travel faster than light, disallowed by relativity, could he get out of black holes.

Einstein's view of spacetime implies timelessness. Does God exist in this way? If so we cannot use temporal terms in connection with him. He is 'outside' both time and space and we have the same problem as with Newton, for how can he enter it? When we think, plan, converse or regret we are engaged in temporal activity but if God is timeless how can he do this? Time and change go together but God cannot change as Plato, Aristotle, Augustine and Aquinas insist. The Greeks thought that change meant decay and imperfection hence the Unmoved Mover could not be associated with it. They believed that a timeless God would know everything, for all times would be present to him and time itself was created with the universe and did not exist in any sense 'before' it.

We might say that God acts timelessly but this has to meet the objection that act and time go together. If there is no time frame in God how can creation take place? When would God make the decision? A timeless God seems to harmonise more with the Steady State theory where a universe always existed but not with creation as an act in the past. The Steady State theory was held by Hermann Bondi, Thomas Gold and Fred Hoyle. The latter was upset when the opposing theory was put forward and ironically give it the name: the Big Bang. Hoyle was to be Stephen Hawking's supervisor in his study for a PhD but he selected another student, Jayant Narlikar, and set him to work on calculations in support of the Steady State theory. Hawking got to know both Narlikar and the calculations! Consequently, when Hoyle delivered a lecture on the Steady State theory, Hawking, to the surprise of those present, queried the calculations saying that he had found errors. Hoyle was annoyed but Hawking was right and became known in a wider circle.[18] It is significant that those influenced by the Steady State theory speak of God as sustainer rather than creator or argue that it does not matter which theory is correct for the doctrine of creation means our dependence on God.[19] But we will present reasons in a later chapter for preferring to think of God as everlasting rather than timeless.

Since time and space are relative there does seem to be more flexibility than in the absolute time and space of Newton. Einstein has replaced Newton's necessary and mechanical order where bodies moved separately and acted on one another instantaneously through empty and uniform space with fields of force. There is a web of relationships with spacetime interacting with matter and integrating everything within it: electricity, magnetism and light. Instead of a Newtonian mathematical framework which imprisons the world within it there is more openness and laws which are not immutable. Time is no longer external but internal and integrated into the process.[20] But the openness cannot be over-stressed for relativity is still deterministic in character. It is true that not everyone agrees about the values of distance and time but the relationships are determined by the mathematics of relativity physics. Reality is viewed as interacting events, not unchanging substances, for matter, motion, space, time, gravity, event, observer, past, present and future all exist in a relationship. With Newton the observer was detached but now we know there is an interconnectedness and interdependence.

The new view of the universe is that it is finite and curved. Thus if we set out from the earth into space in one direction we would return eventually from the opposite direction. Objects that we once thought had objective properties of length, mass, velocity and time are relative to the observer. Can we imply from this that everything is relative including beliefs and values? The question can be applied to the frame of reference of those who wrote the scriptures of the various faiths and, if it is seen as relative to them but not to us, how valid is it? We can react to these questions in different ways. Relativity, with its view of spacetime, applies to travel at high speed and hardly affects ordinary travel. It is Newtonian science that still applies to our everyday world and Einstein did not intend his theory to have such cultural and psychological consequences. It is also necessary to realise that new absolutes have been introduced, the velocity of light:

> ... the spacetime interval between two events is the same for all observers. Everyone carries their own clock and their own time zone, but the order of causally related events does not change ... Einstein took pains to show that while phenomena do vary among frames of reference, the laws of physics are invariant among them. There is a core of relationships which is observer-dependent, though it is described from multiple points of view.[21]

TIME TRAVEL

If space and time are so interconnected, why not travel in time? We can move back and forwards in space so why could we not go back in time like the science fiction writers imagine:

> There was a young lady called Bright,
> Whose speed was faster than light,
> She went out one day in a relative way,
> And returned the following night!

Stephen Hawking is willing to speculate about this. He thinks that time travel is possible and imagines that one day we will be able to make journeys back in time or to the outer reaches of the universe which are 20 billion light years away. The mathematical theory is that if time slows down as a body approaches the speed of light, then once it exceeds it, time would start to run backwards. Imaginary, faster-than-light particles are called tachyons but there is still no evidence that they exist.[22] Einstein was opposed to such a suggestion, saying in 1928 that it was impossible, but in the light of quantum physics and Godel's paper of 1949 he became less certain.[23]

THE METHOD OF SCIENCE

The work of Einstein enlarges the view of how the scientist operates. He found room for imagination, intuition and insight which resembles the mystics and prophets of the great religions. Science proceeds by observation and experiment and postulates theories. Einstein was aware of the importance of observation but engaged in thought experiments which could not at the time be demonstrated empirically. Hawking insists that he does not know of any major theory that has been put forward simply on the basis of experiment. Observations test the theory but the theory is often advanced by the desire for an elegant and consistent mathematical model.[24] Einstein seemed to be able to grasp intuitively a theory but it had to be judged by logical or geometrical reason. He had an artistic sense, aiming for simplicity and beauty, and has rightly been called the artist of science.[25]

GOD

Einstein said that the most incomprehensible thing about the universe is that it is comprehensible. Our mind is in tune with the universe so that we can discover its laws and these can be expressed in mathematical formulae which would seem to point to a mathematical Mind behind it. Einstein believed that in every true searcher of nature there is a kind of religious reverence, for 'he finds it impossible to imagine that he is the first to have thought out the exceedingly delicate threads that connect his perceptions.'[26] He was awed by the mystery of the universe and commented that a complete solution receded as we advanced. There is:

> A knowledge of the existence of something we cannot penetrate, of the manifestations of the profoundest reason and the most radiant beauty, which are only accessible to our

reason in their most elementary forms – it is this knowledge and this emotion that constitute the truly religious attitude.[27]

The universe created a feeling of awe in Einstein and harmonises with the experience of Darwin who had a similar feeling about some of the places that he visited. We also recall the wonder of the American astronauts as they travelled through space inspiring them to read from the scriptures. Einstein did not separate religion and science for he believed that the laws of the universe expressed the will of God and revealed the mind of God. He spoke of ideas coming from God and, when attempting to work out the equations in dealing with gravity and the curvature of space, he said that he was trying to know God's thoughts. We are reminded of Kepler thinking God's thoughts after him. Einstein does seem to equate the laws with God: pantheism. If so he is much nearer to Indian thought than the Semitic religions who in various ways think of God as distinct from the world which he has created. He did say that his God was that of Baruch Spinoza (1632–77) who was born in Holland, the son of wealthy Portuguese-Jewish parents. He had views of God which did not please the Jews of the synagogue and was expelled. Forced to wander about Europe looking for work he gained a livelihood by grinding lenses but found time to write books on *Ethics* and a *Treatise on the Improvement of the Understanding*.

Spinoza believed that outside of God there can be nothing: God is all and all is God. He is a single, eternal, infinite, self-caused principle of nature and of all things. He does not have personality or purpose but all things come from him according to strict law. All the ideas in the universe added together constitute the thinking of God. The sum of our thoughts make up God's thoughts, thus God influences the world of thought and that of things.[28] But there seems to be some doubt about Einstein's pantheism since he often referred to God in a personal way: 'the old One', 'the dear Lord'. It is a true insight, for a God who is impersonal and basically unknowable is hardly likely to be of interest to us.

If Einstein's observer gains knowledge of the world through interaction with it then by analogy we obtain knowledge of God through interaction with him. It is true that we can speculate about his existence and ask if the world reveals him but the primary source of knowledge will be if he can commune with us in a personal way. Such religious experience is witnessed to by the many religious traditions over a long period of time.

Relativity may help us to imagine God as omnipresent yet superspatial. Karl Heim wrote about God and selfhood as other 'spaces' and 'in another dimension'. The same set of events can be differently ordered in different spaces and the flexibility of space shows that they can permeate each other without boundaries. We may therefore imagine God who is omnipresent knowing all events instantaneously since the limitation on the speed of transmission of physical signals between distant points would not apply to him. God would be immanent at all points and in all events. Heim does not make a direct inference from science to religion but derives an analogy from relativity for the presence of God in the world.[29]

More recently, John Houghton, stressing that in relativity the crucial point was introducing time as the fourth dimension, suggests that God could be in a fifth, the spiritual. It would mean that he is 'outside' the bounds of the material universe but able to be present in it. His analogy draws on Dorothy L. Sayers who imagines God as the author of the human drama. The audience watches the play unaware of what is going to happen but the author knows all, hence she is 'outside' of it yet can act in the drama if she so wishes. Houghton is aware of the deficiencies of the analogy for the actors read a prepared script. We may add that it does not leave room for the freedom that we are going to argue for later. But scientists do not rule out that there could be more dimensions than four and since we know that dimensions penetrate one another there is no reason why the fifth should not contain the others. Being 'outside' time would mean the ability to see something as a whole rather than concentrating on the parts. Mozart, for example, was able to visualise and hear a piece of music all at once.[30]

CONCLUSION

Einstein stood in the classical tradition in that he thought he was describing the world as it was in itself not simply its behaviour. He believed that the uncertainties which surfaced in quantum mechanics were due to human ignorance and could be resolved. Yet it was his work on light that opened up the way for the quantum theory. He had not been content with working on the special theory of relativity and in the same year that he wrote his paper on it he also revealed his investigations into the photoelectric effect. He had observed that when light fell on certain metals, charged particles were emitted. If the intensity of light was reduced the number of particles emitted diminished but the speed with which each was emitted remained the same. Einstein suggested that the problem would be resolved if we thought of light coming not in variable amounts but packets of a certain size. Hawking writes that it is like saying one cannot buy sugar loose in a supermarket but only in kilogram bags![31] It was Werner Heisenberg, however, who realised the full implications of the photoelectric effect, pointing out that it made it impossible to measure the position of a particle exactly. It was that uncertainty which disturbed Einstein but should it have done? After all it was he who said that the universe is not only queerer than we imagine but queerer than we could imagine!

Einstein lived in turbulent times and had to stand firm against German nationalism and anti-Semitism. His work was denounced by the Nazis, his books burned, and a reward of £3000 put on his head. On being told about it he said: I'm surprised I'm worth that'. While in America he worked to get Jews out of Germany but even in the US he was suspected by the FBI of being too sympathetic to the victims of fascism. In the midst of all these problems he felt depressed during the last years of his life because he had not found a unified field theory. He was aware of his limitations and, like Newton, thought of himself as a child playing with pebbles on the seashore or entering a library with books in different languages none of which were understandable. At times he doubted that we would ever really know what the world was like.[32] I think the next chapter will explain what he meant.

5 What Kind of World is This?

Science has 'explained' nothing; the more we know, the more
fantastic the world becomes and the profounder the
surrounding darkness.

Aldous Huxley

We mentioned in Chapter 1 that many people believe that
science deals in facts, laws and certainties that resolve the mysteries, superstitions and fantasies of religion, but Huxley
asserts that science has explained nothing. His claim is exag-
gerated but as we enter the quantum world and hear about
imaginary time, the difficulty of predicting how the smallest
units of matter behave, black holes and various views of reality
we will understand his bewilderment.

Three theories changed the scientific world view in the
20th century: the special theory of relativity (1905), the
general theory of relativity (1915) and the theory of quantum
mechanics (1926). Einstein opposed the latter but it was his
work that opened up 'the can of worms'! He was awarded the
Nobel Prize in 1921, not for the theory of relativity – considered too abstract and mathematical – but for his work in
quantum physics which showed that sometimes light in its interaction with matter behaves like a wave while at other times
it resembles a particle. Forms of energy come in quantum
packets and electrons behave in an unpredictable way. The
world which had appeared deterministic became uncertain,
novel and exciting, and more open to a creative dynamic
God. It was Heisenberg's (1901–76) uncertainty principle that
aroused Einstein's reaction: 'God does not play with dice'.
But Stephen Hawking says: 'Consideration of particle emis-
sion from black holes would seem to suggest that God not
only plays dice but also sometimes throws them where they
cannot be seen.'[1]

In this chapter we will discuss the difficulty of knowing the
world as it is, the problem of predicting what is going to

happen, complementarity and holism, and the role of the observer. In this context, we will go on to see that the quantum can relate to the beginning of the universe which poses questions about a creator. But a discussion of the anthropic principle seems to point to design, and the uncertainty principle and the chaos theory indicate a more open universe.

THE PRINCIPLE OF UNCERTAINTY

An electron is a particle, a minute portion of matter with a negative charge of electricity. If a scientist looks at the electron under a microscope she focuses light upon it but this act disturbs its direction and speed. She tries to avoid this by using low frequency light but now there is only a poor determination of its position. If she knows where it is she does not know what it is doing and if she knows what it is doing she does not know where it is. We are reminded that mothers sometimes make a similar comment about their sons!

But does this really matter? Are these not the smallest elements of the universe and what effect can they have on us? They are minute but they are the unpredictable particles which form the foundation of the universe and they do affect us. Quantum theory has given us lasers, nuclear power, the electron microscope and the transistor. These particles are the sort of events which go on in the nerves and the brain and in the giant molecules which determine the qualities we inherit.[2] Heisenberg believed that such uncertainty was a built-in feature of the universe but others including Einstein argued that it was due to ignorance or the imperfection of the instruments used. However, uncertainties occur when nothing is being done to disturb the system: '… The unpredictability of the time at which a radioactive atom spontaneously disintegrates, or the time at which an isolated atom makes a transition from an excited state. Quite apart from any measuring process, a system may lose one form of precision and gain another, in the spontaneous diffusion of wave packets.'[3]

In the Newtonian system the movement of matter could be predicted and it was thought that the world could be described as it is. But now it is realised that we can only describe how the world behaves, not as it is in itself.

The molecules in a gas, for example, behave in an unpredictable way after a collision. On a different level, the snooker player tries to cope with the unpredictable collision of billiard balls and insurance companies weigh up probabilities before issuing a life policy. In quantum mechanics the movement and position of particles follow statistical laws which apply to large numbers not to the individual particle. It displays an unpredictability and appears to show an element of freedom to realise possibilities. The motion of a particle or electron is described by a wave function which contains the possibility for either position or momentum.[4] Novelty can arise for the future is not decided by a law of rigid cause and effect. Einstein attempted to escape from the impasse, but Heisenberg, Bohr and Born insisted that the uncertainty principle was inherent in nature and the final truth about the world. Since the probabilities are influenced by what has gone before we have only a weak form of causality which leaves freedom for one of them to be realised. Chance, which we discussed earlier in connection with biology, would be consonant with the uncertainty principle: a method used by the creator to realise the possibilities. It would not be purposeless but the instrument of creating novelty and enabling the universe to realise itself.

COMPLEMENTARITY

If an electron collides with another it behaves like a particle but if it is fired at a screen containing two slits it becomes like a wave and passes through both slits. Hence the remark: 'Elementary particles seem to be waves on Mondays, Wednesdays and Fridays, and particles on Tuesdays, Thursdays and Saturdays.'[5] The descriptive terms exclude one another but both are needed to give a proper understanding. Putting them together there is an approximate idea of reality: the probabilities, not certainties, of the behaviour of the wave particle. Particle and wave are two sides of the one coin, but it is a shock to realise that while there is an understanding of its behaviour we do not know what it is like in itself, that is, its nature. Those who contend that the nature of a creator is so mysterious that we cannot even consider it fail to reckon with the creation. Complementarity raises questions about determinism and freedom. If an electron can reveal an element of freedom, how much more can we? But when we speak of freedom at our level we use the word choice, whereas an electron arrives at a particular location by chance. And, in any case, we can predict the average behaviour of a number of particles in any given situation. But we may say that the quantum world is much more open and unpredictable.

We proceed analogically from the sub-atomic world to the human and from the human to the divine. We cannot infer from one to the other. Complementarity always applies to the same entity and an electron behaves either as a wave or particle: it is not both a wave and a particle.

HOLISM

We mentioned wholes or holism in Chapter 3 when we were discussing how the reductionist works. Let us take a closer look in the light of complementarity. The reductionist does not accept that the whole is more than the parts but the sum of the parts. She concentrates on the parts and how they work together. The anatomist lays bare the bones, muscles and organs and then hands over to the biologist who looks at the individual cells to discover the molecules. The chemist considers each molecule as an arrangement of atoms and the physicist shows that each atom is made up of particles or the electrons which surround a central nucleus. The nuclei are divided into neutrons and protons and an attempt is made to understand what they are. The procedure is necessary. It is dissecting to understand and has proved very successful.

But at each level there are different explanations which do not exclude one another, for we have complementarity. Consider traffic lights. If they fail, an electrician deals with the mechanical defect so that the light resumes its message of stop, get ready, go. Without the mechanism we cannot have the message. But if someone who has never seen traffic lights asks what the message means he will need a different explanation from that of how they work. The meaning is different from the mechanism but we cannot have the meaning without the mechanism: the higher is related to the lower. Design is there, for someone has decided what colours will represent halting and starting and has planned the mechanism accordingly. Traffic lights are not as interesting as a painting but the same applies regarding higher and lower levels. The chemistry of a painting consists of its pigments and varnishes and the use of these requires skill but the meaning will be considered at a different level. As we ascend through the various descriptions of anything – physics, chemistry, biology, psychology – questions of purpose and meaning will be understood at the highest level not the lowest for the arrangement or pattern of the whole conveys meaning. The atoms of our bodies are continually being replaced so that what we are now is different from what we were when we were younger. Yet we remain the same person throughout life: 'the permanence lies in the way the atoms are arranged. It is the arrangement, and that alone, that gives rise to the distinctive person.'[6] Only if science can explain the characteristics of a person on the same level as chemical reactions can it be said that we are sophisticated machines. We will consider the attempt later.

There is also a 'wholeness' about the quantum system itself. Bohr argued against Einstein that we cannot regard particles as isolated from one another. What is done to one affects another. He demonstrated that if we were extracting information from two electrons which were not in physical contact, the system to which they belonged would be affected. They would show in their togetherness laws of behaviour different from the laws which govern them in isolation. An example of this is, 'action at a distance'. Here there is internal relationship not simply external, indicating an undivided wholeness. It was not a case of adding the properties of the two electrons together but taking into account the connectedness of the system: a quality that could not be explained in terms of physical forces. It is something which cannot be identified, which makes things difficult for the reductionist who insists that any concept such as 'spirit', for example, that does not have an identifiable counterpart in the physical is ruled out.[7]

THE OBSERVER

What is going on in the atomic and sub-atomic world cannot be observed directly but only traces of movement. What the observer experiences is related to her frame of reference. She decides by her experiment whether the electron will behave as a wave or particle. Thus in choice of motion relative to what is observed and the decision whether to measure position or velocity the observer is decisive. The quantum object reveals itself on the basis of how she examines it. Quantum theory is a theory of the observation of nature not of nature apart from observation. There is an interrelationship between subject and object, for '... the liquid shivers under the lens ...', and the object is not static and certain, 'the flow of gas is shot through and through with the random darting of its particles'.[8] If we insert a thermometer into a liquid we make the liquid slightly cooler depending on how cool the thermometer was. The act of measurement has changed what we set out to measure, hence it is necessary to heat the thermometer to the same temperature of the liquid before doing the test. But when the object is a microscope, and the electrons involved tiny, light will affect.[9]

Does this concentration on the observer mean that the primary qualities of an object – length, mass, velocity, time – are now related subjectively to her? Is it her mind that is creating reality? One famous philosopher in this tradition was Bishop Berkeley (1685–1753). He taught that there is no physical world in the sense of an independently physical object. What we call physical objects are collections of ideas in the mind. We know the appearances of objects but not their real nature. To exist is to be perceived. If we think this is inconceivable imagine losing all our senses, would the world then exist for us? But Berkeley's students were not impressed and wanted to know if the tree in the quadrangle was still around when Berkeley left it! Berkeley's answer was that it continued to exist in the mind of God but the students wrote the limerick:

There once was a man who said, 'God
Must think it exceedingly odd
If he finds that this tree
Continues to be
When there's no one about in the Quad'

Reply: Dear Sir: Your astonishment's odd
I am always about in the Quad

And that's why the tree
Will continue to be,
Since observed by
Yours faithfully, God.

Berkeley believed, and he has had his successors, that the
world is the creation of the mind of God and therefore is basically mental not material. It is interesting to note that a physicist, Frank Tipler, agrees with him.[10] But it is objected that
measuring devices such as clocks and metre sticks can take the
place of the scientist and what happens can be recorded on an
automatic camera. There is an 'otherness' about the interactions for some of them are not under the control of the observer. He can disturb by his use of instruments but it is the act
of measuring which is the problem not the entry of his mind.
Yet, as the example of Schrödinger's cat shows below, when the
observer himself is directly involved his examination affects the
outcome. Uncertainties also occur when nothing is being done
to disturb so it would seem inherent in nature. The important
point is that we obtain knowledge of the world not in a detached way but by interaction. However, as science has probed
into the shadowy and unreliable world 'at its sub-atomic roots'
the clear and determinate world of Newton has 'dissolved into'
the cloudy fitful world of quantum theory'.[11]

An electron can possess an 'up' or 'down' spin and which
state it is in is not determined until it is looked at. Before the
observation it is potentially in both states but when we look at
it, the electron 'decides' to be in one particular state. If we
think of a particle which may or may not decay within a
certain period of time, until it is examined, it is in a state of
'having decayed' and 'having not decayed' simultaneously
from the observer's point of view. The example of
Schrödinger's cat is famous. A particle is likened to an apparatus which contains a deadly poison. If the particle decays, the
apparatus releases the poison; if it does not decay, the poison
is not released. The apparatus is placed inside a sealed box
with the cat. The observer does not know until he looks
whether the cat is dead or alive. It is all a matter of probability.
But when he examines it, he will affect the particle which
'decides' to release the poison or not. Thus the observer and
what is observed are bound together in the quantum world.[12]

This is difficult to accept, for according to classical physics we can assume the definite state or history of something. But in quantum mechanics the cat can have two histories: one in which it dies and the other where it lives!

This has been called the 'sum over histories' by Richard Feynman. A system does not have a single history in space-time. We assume that a particle at a point A will move in a straight line away from it but according to the sum over histories it can move on any path that starts at A. Hawking says that it is like placing a drop of ink on a piece of blotting paper. The particles of ink will spread through the blotting paper along every possible path. But if particles can take any path it may be possible for them to travel faster than light and thus escape from a black hole. The probability would be even greater with small black holes rather than large ones.[13] We will return to this idea when discussing Hawking's imaginary time.

How do electrons exist? We are inclined to think of them as being like small-scale billiard balls because it is difficult to get away from Newtonian ideas. But these electrons, though real, do not exist in the sense of having well-defined places and speeds. The physicist simply thinks of them in mathematical terms enabling him to relate the results of their behaviour to his experiments. Physicists have realised the importance of the quantum in connection with the beginning of the universe and have tried to create models of creation by combining the principles of quantum mechanics with relativity. It is speculated that quantum fluctuations are the primary cause of the universe which has implications for the religious view of a creator.

THE BEGINNING OF THE UNIVERSE

The Semitic religions teach that the world has a creator and Augustine believed that time was created with the universe and was a property of it. Scientists thought that the universe was static, until Hubble in 1929 pointed out that the galaxies are moving away from us:

> The situation is rather like a balloon with a number of spots painted on it being steadily blown up. As the balloon

expands, the distance between any two spots increases, but there is no spot that can be said to be the centre of the expansion. Moreover, the farther apart the spots are, the faster they will be moving apart.[14]

This expansion from the Big Bang, the singular event of infinite density from which the observable universe appears to have originated approximately 15 billion years ago, can be explained by the inflation theory: the size of the earth has kept doubling at a tremendous rate. But there is debate about the rate of such expansion which would leave the universe in a non-uniform state.[15] At least one model of the inflation has failed but another called the chaotic inflationary model appears better. If it could be proved, it would eliminate the need to think of the initial state of the universe being chosen with care. But it does not explain how there was a mechanism in the first place for a fine tuning expansion.[16] The mass density of the universe is perfectly tuned to the critical value from its earliest expansion. If the mass density had been greater than the critical value, it would have moved to a Big Crunch or if the mass density had been less than the critical value, gravity would not have been able to stop the expansion. The amazing thing is that mass density is equal to the critical value so that it expands just fast enough not to collapse.[17] There was a delicate balance between the law which we can regard as the pattern and strength of the fundamental forces of nature and circumstances seen as the particular way in which the world emerged from the Big Bang. Science cannot give an explanation of its own laws, law and circumstances are assumed. But the scenario points to design, hence the theist argues that the laws are God given.

THE ANTHROPIC PRINCIPLE

Hawking paraphrases the anthropic principle as meaning: 'Things are as they are because we are.'[18] The possibility of life on this planet depended upon a delicate balance of the basic forces of nature and on specific initial conditions. It appears that we were meant to be here; 'anthropic' is derived from the Greek word anthropos meaning man. Steven Weinberg believes that quantum cosmology provides a context in which the anthropic principle becomes simple common sense. The most probable universe is the one which we are in: it is the best of all possible worlds.[19]

The ingredients needed to provide a universe where life would emerge would require carbon and, since none came out of the Big Bang, it probably came from the stars. An explosion there would drive out the carbon and oxygen. If this was the case, a fine tuning of a gravitational force would have been needed to generate an explosive blast of neutrinos. But how did the stars originate? We can think of two forces: the Big Bang and the gravity force. If the gravitational force had been too high then the universe would have disappeared but if the expansion rate had been too great then the universe would have expanded at such a rate that gravity would have been unable to form stars and galaxies. These forces would have to be balanced with amazing accuracy in order to form stars and galaxies so that eventually carbon would be available for the evolution of life. Hawking has said that if one considers the possible constants and laws the odds against a universe that has produced life like ours are immense. Consequently, it is difficult to believe that there was no intention or design behind it for without such a universe evolution could never have started.

The weak anthropic principle states that in a universe which is large or infinite in space and/or time, the conditions necessary for the development of life will be met only in regions that are limited in space and time. It must be in a state that will enable life to develop. But this weak form does not tell us much. The strong form of the principle states that there are many universes, that is, the interpretation of quantum theory that supposes that all the possible outcomes of a measurement in fact occur in parallel disconnected worlds into which physical reality divides at each act of measurement. It was only by chance or accident that the right conditions occurred here and life developed, so a Designer is not required. But how do we know that these universes actually exist? And would they have life? This speculation introduces a complexity to explain the regularities of our universe but the scientist likes simplicity.

It is possible that he may in the future be able to give an explanation of these initial conditions but currently they imply design. It is the design of the universe, not the evolution of man, so it is immune from Darwinian assault. Paul Davies points out that Darwin's theory required a collection of individuals or organisms upon which natural selection can operate. His example is of polar bears. How did they come to blend so well with the snow, thus having an advantage in the survival game? We can imagine a collection of brown bears hunting for food in a snowy land where their prey can spot them and escape easily. The bears go hungry! Then by some genetic accident, a brown bear gives birth to a white bear. It creeps up on its prey and doesn't go hungry. It lives longer and produces more white offspring who fare better and multiply. They take all the food and drive the brown bears to extinction. But the argument depends on selection from a group. However, with the initial cosmological conditions there are no competitors, for such laws and conditions are unique. Davies rejects the many universes theory. Having considered the arguments against design he concludes that the present understanding of physics cannot decide against it. The question remains open, but for his part he cannot believe that he is here by some accident.

The most extensive discussion of the anthropic principle has been undertaken by Frank Tipler and John Barrow. The structure of the universe is such that life was always a possibility. Tipler says that the strong anthropic principle cannot be disproved and that Fred Hoyle accepted it. Both think that the many universes view places too much stress on accident.[20] What we can say is that the principle does not give us proof of the existence of God but adds to the cumulative evidence, which we will discuss later. It also points to the value of mankind in the divine scheme of things. The earth, removed from the centre of things by the work of Copernicus, regains its importance as the place where the conditions were right for life to begin.

NO BOUNDARY PROPOSAL

Most physicists continue to hold that the Big Bang was an expansion from a singularity which is an edge or boundary to spacetime. Some think that a singularity is the nearest thing that science has found to a supernatural agent.[21] The theory of general relativity cannot tell us much about it being unable to cope with the infinite curvature of spacetime and the infinite density. All the known laws of science break down at this point for mathematics cannot deal with infinite numbers. But we are dealing with the very small, hence quantum effects are important. Hawking sees the need to use a quantum theory of gravity and dispense with singularities. If it could be done, then quantum laws would hold everywhere and apply to the beginning of time. But so far there is no consistent theory that combines quantum mechanics and gravity. It is Hawking's ambition to find one.

Earlier, Hawking and Penrose thought that there was a singularity at the Big Bang with the universe existing in a hot, dense state. Hawking concentrated on black holes which are voids in space caused by the collapse of a star due to loss of fuel. Not even light can escape from black holes but they collapse when they reach a singularity of infinite density and spacetime curvature. Hence the connection with the Big Bang at the beginning of time. Black holes are another example of mathematical models being developed before observations could prove them correct.[22] They can be detected because they exercise a gravitational force on nearby objects. It is thought that many of these existed at the beginning of the universe so understanding them will help us to know about its early stages. But, despite their name, they do glow like a hot body and the smaller they are the more they glow. Black holes may not be black after all.

In 1974 Hawking discovered that the black holes in which a singularity occurs radiate and possess temperature and entropy. He came to see that according to the quantum theory these objects would emit X-rays and gamma rays which cause the black hole to evaporate slowly but what happens to it is unknown. There is the collapse from a low-density state to one of high-density: the reverse of the expansion of the universe which moves from high to lower densities. Black holes are known only by their effects. John Wheeler in this connection says we should think of a dance with the boys dressed in their black evening tuxedos and girls in white dresses whirling around in each other's arms. The lights are turned down low so that all you can see are the girls. The girl is the visible star and the boy is the black hole. You cannot see the boy but the girl whirling around provides convincing evidence that there is something holding her in orbit. The X-ray satellite, Uhuru, launched in 1970, has discovered over 300 sources of X-rays. One of these is in the constellation Cygnus which is orbited by a blue star. It had to be a black hole for it was too massive to be a neutron star.[23]

Hawking records that in November 1970, shortly after the birth of his daughter Lucy, he was thinking about black holes as he was getting into bed. Suddenly he had a flashing insight: the surface area of a black hole can only stay the same or increase but never decrease. If an object has temperature it has thermal radiation hence black holes were not completely black. At the surface of the black hole particles can be attracted by the intense gravity. Two particles make up a virtual pair, one negative and the other positive. What happens at the surface of the black hole is that the negative one is attracted into the hole but the positive one escapes in the form of radiation.[24] Thus black holes did obey the laws of thermodynamics like everything else in the universe. Particles could escape from black holes and continue their histories.

Hawking based his view on the quantum principle that one could avoid histories having a beginning in time at the Big Bang. Hence he thought of sum over histories and imaginary time. He argues that a particle does not have a single history because it follows every possible path in spacetime. It travels in real time and imaginary time. One can add up the probabilities for particle histories with certain properties, that is, passing through certain points at certain times, and then extrapolate the result back to real spacetime in which we live.[25] He speaks of imaginary numbers by which the distinction between time and space disappears and there is no difference between time direction and directions in space. If we draw a horizontal line going from left to right on a sheet of paper with early times on the left and late times on the right we have ordinary time. But we can also draw another direction of time, up and down the page. It is the imaginary direction of time, at right angles to real time.

But why introduce imaginary time? Because matter makes spacetime curve in on itself leading to singularities where spacetime comes to an end. The equations of physics do not work and we cannot predict what will happen. The imaginary time direction behaves in a similar way to the real time but results in a spacetime that closes in upon itself, without boundaries or edges. It would not have any point that could be called a beginning or an end and the laws of physics could operate. If we knew the history of the universe in imaginary time we could calculate how it behaves in real time and get the unified theory. Hawking is saying that what applies to a particle could be applied to the fabric of spacetime but he admits that we do not as yet know how to do this summation.[26] But imaginary time may not be good terminology since it could be equated with the way we use the word imaginary in everyday language. Hawking notes the objection and records the attack of the philosopher of science who asked: 'How can a mathematical trick like imaginary time have anything to do with the real universe?' But mathematicians do use imaginary time, it is a well defined mathematical concept.[27]

Hawking, then, has moved to the no boundary proposal in cooperation with Jim Hartle. This proposal suggests that space and time are finite in extent but have been closed up on themselves without boundaries or edges. It means that there would be no singularities and the laws of science would hold everywhere even at the beginning of the universe, but there would be no beginning and no moment of creation.[28] What Hawking tries to do is to apply quantum cosmology to the universe 'before' the Big Bang and determine its wave function. We have seen that an electron can behave as a wave or a particle and Max Born proposed that the wave function around the nucleus of an atom was a probability wave. One cannot say exactly where it is located. If we assign cosmological models of the universe a wave function we can think of possible universes and probable ones. Then by choosing universes with no boundaries it is possible to see results consistent with ours. In real time they have a beginning and end but not in Hawking's imaginary time. Quantum fluctuations (seeds for galaxy formation) play an important part in the forming of the universe. In 1989, Cosmic Background Explorer measured the Big Bang background and the variations showed that quantum fluctuations are present in an inflationary universe and consistent with Hawking's no boundary proposal.[29] It has been noted that such imaginary time has a parallel in Plato and Augustine.[30] Augustine believed that the world and time were created together and the world was an idea in the mind of God from eternity, and Plato insisted that perfect forms or ideas had their copies here. Hawking's imaginary time could be viewed as the perfect idea with our time an illusion. Einstein had regarded it in that way.

If Hawking is right there would be no need for a creator. Before Hawking's proposal it could be argued, as we have done, that the universe evolves according to laws which stem from God. He set it going in the way he wanted and the present state would be the result of his choice of the initial conditions. But in the no boundary proposal the laws would hold at the beginning of the universe and God would not have the freedom to choose the laws that it obeyed. Hawking notes this and comments that any choice God had would be limited. But:

> ... even if there is only one unique set of possible laws, it is only a set of equations. What is it that breathes fire into the equations and makes a universe for them to govern? Is the ultimate unified theory so compelling that it brings about its own existence? Although science may solve the problem of how the universe began, it cannot answer the question: why does the universe bother to exist? I don't know the answer to that. If we did we would know the mind of God.[31]

Hawking insists that what he is proposing says nothing about whether God exists or not. His proposal means that we would not have to say that God chose to set the universe going in some arbitrary way which would be difficult to understand: 'It says nothing about whether or not God exists – just that he is not arbitrary'.[32] He seems to think that our usual view is that God initiated the universe as a personal whim. Perhaps he has in mind the Hindu thought of Brahman bringing the world into being as a kind of sport. But the Christian belief is that it was because of his love that God created the world and engaged in the work of reconciliation. Hawking believes that we still have the question: why does the universe bother to exist? And he does not rule out that God can be the answer to that question. He was not prepared to accept the advice of the pope to leave what happened 'prior' to the Big Bang to religion but proceeded to get rid of the singularity by his imaginary time. Paul Davies queries his proposal because it is difficult to apply a theory of sub-atomic particles to the entire cosmos. It is usually applied to the behaviour of component parts and it is difficult to explain how a relatively non-quantum world emerged out of a quantum.[33] It has also been pointed out that quantum fluctuations occur at certain positions in spacetime but there was no spacetime to begin with. Science, as Hawking notes, cannot answer the why question regarding the universe because it involves motive and intention and these are important for any discussion of why something has occurred.

Hawking seems to accept the weak form of the anthropic principle but his supergravity theory tries to explain the initial conditions. Apart from his work there are the superstring theories where the basic objects are not particles but extended objects like little loops of string. Particles are like vibrations on a loop but, as Hawking says, there has been little success in obtaining experimentally testable predictions from such a theory. There is a religious tone in his writing and he contends that we cannot discuss the beginning of the universe without reference to the deity. But his God, like that of Einstein, is impersonal. With regard to the anthropic principle many believers insist that it confirms their faith that God is Creator and Designer of the universe. He brings order out of chaos (Gen.1.1ff). According to Genesis the earth is as it is because God intended it for us and he determined the conditions.

CHAOS AND OPENNESS

Quantum mechanics shows that initial conditions involving the very small are difficult to predict because of the uncertainty principle. It is a problem when trying to understand the beginning of the universe, but when we add the revolutionary chaos theory, the world appears even more open and unpredictable. In this theory, chaos does not mean unrestricted and haphazard behaviour but a sensitivity to circumstances which affects a system. It can react back on itself and explore a range of possibilities: a 'stranger attractor'. It exhibits a kind of ordered disorder. An example is the weather. The last hurricane in England took the forecasters completely by surprise because some slight sensitivity affected the system. Clouds are not clocks. Yet such systems displaying freedom of behaviour are able to arrange themselves into a order of a kind: patterns formed by clouds. Order emerges from chaos.

Ilya Prigogine, the Russian chemist, drew attention to open systems and won the Nobel Prize for his work. Newton had concentrated on systems which were stable, orderly and uniform, and displayed equilibrium. A body remains in a changeless state when it has no forces acting on it or if they do, it achieves a balance. There is no change or loss of energy for it is a closed system. The claim of Prigogine and his colleagues is that while some parts of the universe do operate like this they form only a small part of it. There are open systems which exchange energy or matter or information with their environment. Fluctuations upset previous organisation and occur at a bifurcation point which is a critical moment making it impossible to determine in what direction change will occur. The system may disintegrate into chaos or find a new higher level of order and organisation. Alvin Toffler takes an example from society. A tribe's birth and death rate may be in equilibrium so that its population remains stable but if there is a sudden rise in the one and not the other the equilibrium is upset. If the birth rate soars the tribe is disturbed for food and other resources are not available. The result is that the entire system has to reorganise itself and it may do it 'in ways that strike us as bizarre'.[34] The same would apply to the individual where the shock of sudden change may lead to better organisation of life or have a deteriorating effect. A bifurcation point could be seen as a point of choice making it impossible to tell what will be the outcome.

We are open systems in that we interact and exchange information with the environment. When we try to close ourselves off from others we become introverted and often distorted in our behaviour. If we are open to others it is likely that we will be open to God which may be the meaning of the image of God in which we were made (Gen.1.26). It is one of the ways that man is different. As the Hindus say: 'man is the only animal who offers sacrifices'. He is distinctive not only in worship but in his mental capacity which operates at the higher level of his being. The sensitivity and unpredictability of complex and dynamic systems enhances our view of cause and effect since there is more flexibility.

Various connections between religion and the quantum world have been made. One has been the work of Fritjof Capra with his book, *The Tao of Physics* (1975). He sees parallels between the picture of the world described by quantum theory and Eastern religion. Capra's mysticism builds on molecules and atoms vibrating together, and millions of particles coming into existence and disappearing again. It made him think of the atoms of his body participating in the cosmic dance of energy. It happens as protons and electrons rearrange themselves into combinations: a kind of electric dance of whirling electrons and protons. Capra thought of the Hindu dancing god Shiva, one of a triad with Brahma and Vishnu, and felt it was like the rhythm of the sub-atomic world. As Lord of the Dance he dances out the creation of the world. The idea is also reflected in a popular Christian hymn composed by Sydney Carter which makes Christ the Lord of the Dance:

I danced in the morning when the world was begun,
and I danced in the moon and the stars and the sun,
and I came down from heaven and I danced on the earth;
at Bethlehem I had my birth.

When Shiva gets tired of the dance he relapses into inactivity and the cosmos becomes chaos. The myth of Shiva is more terrifying than joyous. But the thought of creation as a dance is significant for there is order and movement or necessity and chance. Capra believes that both reason and intuition have a place in science and that models of what we are like and the world play a part in both. Religion often begins with observation as seen in Taoism, that is, kuan (to look), as people encountered prophets and the Founders of the Faiths. But he says that visual perception is soon transcended by non-sensory experience of reality: mysticism.

Mysticism is not ruled out by Bertrand Russell: ... 'Even the cautious and patient investigation of truth by science, which seems the very antithesis of the mystic's swift certainty, may be fostered and nourished by that very spirit of reverence in which mysticism lives and moves.'[35] But Capra may not be giving enough place to reason. The rational framework of the scientist does not correspond with his concept of the mystic's consciousness where a person's individuality dissolves into an undifferentiated oneness that transcends the world of sense and leads to the disappearing of the notion of 'things'. Physicists do recognise intuition and insight but it seems to occur in a context: a mathematical problem or an experiment or when Darwin was reading Malthus. The mind is full but in mediation the attempt is made to empty the consciousness of all concepts and ideas.

Capra has been criticised for holding the outdated 'bootstrap' concept of matter. It is based on a particle being the result of a coming together of other particles and its exchange between these can create the force that holds them together to make itself. It can be the cause of its own effect! This encouraged the belief that there are no basic elementary particles but everything is made out of everything else in an act of self-consistency. The world has been pulled into existence by its own bootstraps. But the more orthodox view prevails that matter is composed of identifiable constituents: quarks and gluons, and their properties are manifested by symmetrical relationships.[36]

CONCLUSION

In preceding chapters we noted that classical physics was inclined to stress order, pattern and law, with a static nature. In Chapters 2 and 3 the biological work indicated development or becoming. Now in this chapter we have chaos or lack of order and novelty, unpredictability and creativity but balanced by statistical law. There is chance and necessity. Hence if the world reflects the creator then he is a God of novelty and surprises. We can also imply that if the world exhibits a freedom it has been given to it by the creator and he allows the tendencies in matter to realise themselves.

Relationships are stressed with particles affecting one another even at a distance. An electron moves from one orbit to another when it is disturbed by a unit of light energy called a photon. It has now more energy and moves into a different energy state. These transitions cannot be predicted with certainty. We need to think of one event related to another rather than cause and effect. We saw in the last chapter that causes from behind were not a sufficient explanation of how things happened because we had goals which we wanted to attain in the future. Holism is confirmed by physics and its principles of a pattern-forming kind act downwards as a cause. It is seen in willing and thinking which control actions and there may be holistic laws of nature driving the evolution of complexity, at present unknown to us but in principle discoverable by science. It also seems entirely coherent to suppose that God interacts with creation in the same way. God is not manipulating systems but informing by revelation through his spirit. Perhaps there is some likeness here to the 'guiding wave' in quantum mechanics which influences the motion of a particle. However that may be, there is a sensitivity and openness about the systems under discussion and we see that in ourselves. Isolation and failure to relate properly to others result in a disintegration of personality.

The observer is related to nature in a way that Newton could not have foreseen, affecting by his measuring what he wants to understand. He interacts with what he is doing. In the past the action of God has often been seen in an interventionist and direct way and even in a scientific age when lightning strikes a cathedral the media declares it may be God's way of telling us that he does not want a 'heretical' bishop! But interaction with the world which has its own freedom is a better way to think of God's relation. We have also discovered in the study of the quantum that we can only have partial representation of what the world is like though there is no reason to doubt that it is real knowledge. We are also limited in our knowledge of the nature of God but we will argue that revelation gives us true knowledge in accord with our ability to receive it.

We have been considering the mysterious world in which we live but now we turn to the crown of evolution, mankind, to see if we can discover what we are like.

6 What are We?

Because of the prestige of science as a source of power and because of the general neglect of philosophy, the popular Weltanschauung (world view) of our times contains a large element of what may be called 'nothing-but' thinking. Human beings ... are nothing but bodies, animals, even machines ... values are nothing but illusions that have somehow got themselves mixed up with our experience of the world; mental happenings are nothing but epiphenomena ... spirituality is nothing but ... and so on.

Weizenbaum

Weizenbaum is summarising the programme of the reductionist scientist who insists that we are nothing but atoms, molecules, cells and organs, and these can be chemically analysed and mechanically explained. As Greta Garbo said, love is only a chemical reaction! This is an incomplete explanation and does not explain the human characteristics that distinguish us from animals and machines. In order to argue for such a distinction we will consider in this chapter computers, animals, mind and brain, the concept of a person, the self, free-will, values and our religious propensity.

THE COMPUTER ANALOGY

In the computer analogy, the mind – information processing, organisation and patterns – equates with the computer software, and the neurons of the brain are the hardware. The programs are in the genes. But is the human memory analogous with that of a computer? It can be located in a computer but where is it in the brain? There is disagreement about where memory traces are: local or non-local? Portions of the cortex can be removed yet memory remains which implies that it is global. Brain damage can cause amnesia but memory will return. Amnesia has many causes and not all of them are to be equated with loss of memory in a computer which cannot recall its memory traces. Human memory appears distinctive, involving recall and perception, it 'is able to perceive a new order or a new structure, that is not just a modification of what is already known.'[1]

Experiments have been carried out to remove disturbing memories and implant new ones. Electroconvulsive therapy is used with patients suffering from severe depressive illness but heavy doses reduce them to the level of a vegetable. American prisoners were subjected by the enemy to 'brain washing' and loyalties removed so that they denounced their country's war policy on TV. It is asserted that the CIA tried similar techniques to remove memory traces and create amoral assassins but failed to combat the mind's opposition. Today the hope is to make a machine that will replicate human intelligence. If one region of the brain is damaged it is possible for other parts to take over the function. Unlike computers, brains are plastic and can remould when injured. Continual replacement of molecules and cells occurs in our bodies yet memories remain. No computer could survive such a complete turnover of the constituent parts.

Computers are excellent at mechanical operations and far exceed human capability, but they are automatic. The nearest parallel is an automatic action or the working of our bodily organs. Neurons in the brain are not the same as chips, circuits and so on, in a computer, for they are indeterminate. Hence Roger Penrose's contention that there is a built-in indeterminacy in neural responses. We cannot understand the mind or brain by analysis of individual components because their responses are inherently not predictable. But such indeterminacy at the level of neurons gives way to predictability at the level of the system. Thus the mind is a global attribute of the brain and is not a local product of any of its parts.

A computer can simulate emotions but simulation is not the real thing. Computers are so well programmed to play chess that they have beaten the masters. But it does not mean that computers are aware of what they are doing so that the unexpected can result in failure. This happened to the supercomputer 'Deep Thought', which could not report why it had made its blunder although the programmers knew. Computers compute but the mind has an awareness, insight and creativity that programs the mechanical computation.

After an exhaustive survey of computers, minds and the laws of physics, Roger Penrose concludes 'that mere computation cannot evoke pleasure or pain or poetry or what we feel looking at the beauty of an evening sky or the magic of sounds or hope or love or despair.' Consciousness is not something which has just 'accidentally' been conjured up by a complicated computation for 'it is the phenomenon whereby the universe's very existence is made known'.[2]

Brains and the organisms which they inhabit are not closed systems but in continual interaction with the world. Unlike computers, brains are not error-free machines, they are capable of modifying their structure, and have the ability to make judgements. A computer is a dead storage system but the brain is living and dynamic. Memories are stored but the brain has a capacity to connect and relate one thing to another in a way that a computer does not.[3] But we must not underestimate how computer science is advancing. Joseph Weizenbaum wrote a program called Eliza for a computer so that it could act as a psychologist. He wanted to show the limitation of the computer in such a dialogue but some people seem to prefer pouring out their problems to it rather than their psychologist. When used in hospital one old lady said: 'I've never had someone take so much time with me before'![4] It is asserted that the biological computer will be a new form of life: an evolutionary leap. Turing, Michie, von Neumann, the pioneers of such a development, envisage robots doing domestic and hospital work, fire-fighting, mining and providing companionship for us. There would be many benefits provided that they did not develop into modern Frankensteins.

But it is difficult to construct an artifical brain with the characteristics of the human. A machine needs to be programmed for calculation and rules established: the construction of algorithms. Algorithms are ways of finding a solution to a problem by testing every alternative in order, until the correct response is produced and verified. The machine will work according to these rules but with problems that require trial and error or 'rule of thumb' there are difficulties. The machine cannot judge its own program except to respond that it cannot carry out the task. It is the human that supplies the program and judges the applicability of the rules for any given situation.

This judging or 'seeing', according to Roger Penrose, is the essence of mind and is essential to mathematical argument.

Weizenbaum points out the limitations of the most advanced computer to acquire information without being 'spoon fed'. Not all human knowledge is encoded in information structures for there are other ways of communication as when one human expresses emotion by touching the hand of another. This involves at the very least having a human hand! No organism that does not have a human body can 'know' in the same way that humans do. Language used by humans is more than functional, unlike a computer language, with which a computer identifies things and words only with immediate goals to be achieved or with objects to be transformed. Our language identifies hopes and fears and love, reflecting a socialisation that the computer cannot know. The making of a human person requires a socialisation from the cradle to the grave and involves a struggle between one's outer and inner world. Man's 'life is full of risks, but risks he has the courage to accept, because, like the explorer, he learns to trust his own capacities to endure, to overcome. What could it mean to speak of risk, courage, trust, endurance and overcoming when one speaks of machines?'.[5] If the arguments of Penrose and Weizenbaum are right computers will always be subject to such limitations. We are subjects of experience, more than information processing, possessing creative thinking, feeling, consciousness, self-awareness, which sets us apart from machines.

THE HUMAN AND THE ANIMALS

Animals have desires and can adapt to their environment and instinctively prepare for the future. They learn and form relationships, but our mental ability operates on a higher level with a complexity of motives, intentions and beliefs. At the lower levels of animal life actions are automatic, based on instincts and reflexes, whereas we have a cognitive system and a much more developed brain. The animal has little consciousness that the regularity may be disturbed. The chicken, as Bertrand Russell said, expects to be fed every day and has no awareness that one day its neck will be wrung. We have greater capacity than the animals to understand sensory data, we can mentally manipulate ideas and plan ahead. We create as well as respond to our environment.[6] Animals communicate with one another through signs and signals but understanding is rudimentary. There is a humorous story of a man living in a totalitarian country who taught a parrot to speak and when it escaped from the cage reported its loss to the police. He described its colour, size, shape of beak and so on. The policeman asked: 'And does he say anything?' 'Well, yes,' replied the man; and then hastened to add: 'but his political opinions are entirely his own'![7] Parrots can be taught to count and recognise shapes and colours and indicate when they want something and most animals can learn by stimulus, reward, reinforcement, trial and error, but as Thorndike demonstrated with cats most of them have little understanding of what they are doing.[8] We cannot ascribe understanding or beliefs to the sub-human world but we can state our beliefs and give reasons for them or decide not to reveal them.

The development of language is essential to the human and at around 18 months the average infant will be able to combine words into two word sentences. Skinner held that language is learned while Chomsky contended that the ability to combine words and the desire to communicate was innate. The best view appears to be a balance between the genetic and environmental elements. We are genetically programmed to use language but the particular language we employ depends on the one we hear and imitate.[9] It is only with maturation that abstract thinking develops where we can combine ideas and what we observe. Some hold that language determines thought so if a language does not contain words for speed then we cannot think of doing something faster than someone else. If we do not have words to describe various colours we cannot perceive differences between them. Psychologists call it the linguistic relativity hypothesis since it means that some cultures have ideas that cannot be translated into others. But most ideas can be translated into another language and languages continually change to incorporate words from other cultures. While admitting that the structure of the language will influence the way we see things it is possible to perceive something without knowing the word for it. Babies and young children do this all the time. What is likely is that language makes us think about something in a certain way but it does not determine these thoughts and perceptions. According to Piaget, language skills develop when the child is cognitively mature enough to understand so language reflects thought. It is the use of language which has had such effect on cultural evolution.

But what about chimpanzees who are most like us?[10] Chimps can be trained to do tasks and show signs of self-awareness when they are allowed to play with mirrors. First attempts to teach a female chimp called Washoe to 'speak' in sign language were disappointing. But Sue Savage-Rumbaugh, a researcher, continued the program with Sherman and Austin, a pair of chimps, who used language to listen to one another and cooperate. She reared pygmy chimps (bonobos) and they acquired understanding of language at least equal to a three-year-old. Further experiments and observations convinced her that apes showed the clearest evidence of conscious thought among non-human animals. On the basis of these and other investigations it is possible to credit animals with awareness, rudimentary elements of thinking, the aiming at goals and considerable ingenuity in dealing with obstacles. Chimps can learn to use signs and symbols meaningfully and accurately but the uniqueness of the human is shown in his or her ability and creativeness displayed in combining words into an unlimited number of sentences.[11] The development of mind involves full consciousness of past, present and future. Information is received and transmitted in a sophisticated way and there is motive and intention which we cannot predicate of the animals. While there is similarity in various ways between the human and sub-human there is a vast difference of degree.

BRAIN AND MIND

Philosophers have adopted various positions regarding the relation of body and mind. After spending a day doubting, Descartes separated mind and body into distinct substances. He asked himself the question: Is there anything left that I cannot doubt? Of course, he could not doubt that he had spent the day doubting or thinking! 'I think,' said Descartes, 'therefore I am or I exist' (*cogito, ergo sum*). The mind was the free director of a mechanical body and since animals did not think, they were machines. But this view contradicts the biological description of mind evolving from matter and the presence of some rudimentary form of consciousness in sub-human life.

The psychological behavourist goes to the other extreme and says that thinking is only a bodily response to stimuli. We can tell what is going on in the mind by observing actions. This can be explained mechanistically by cause and effect. Skinner experimented with rats in boxes, watched their performance, and then applied it to the human. It has been called the fallacy of ratomorphism in that the behaviour of the sub-human is projected onto the human. But we can never be certain that experiments carried out on animals will yield the same result when they are applied to us for the mechanisms involved in the behaviour of one species may not be the same as that of another.[12]

It is a common fault that we attribute human-like qualities to dogs or cats, that is anthropomorphism. The same applies when we try to make God in our image. But we can learn a lot about people by their behaviour. If someone behaves rudely we excuse it by anger or being up too late or … but if the person's behaviour is consistently bad we say that it is the result of their character which is difficult to penetrate. Observation can only go so far. I see someone in a darkened room lifting a glass to her lips. What is her intention? Is the drink to quench thirst or is it for medicinal purposes (a woman told me that she always drank gin for that reason!) or is it an attempt at suicide? I cannot know simply by watching actions for they can conceal feelings. We show 'a brave face' when suffering from pain or receiving bad news. Thus we have to move from observation to make inferences concerning intention and motive.[13]

Gilbert Ryle in *The Concept of Mind* called the mind the 'ghost in the machine'. We consider 'thinking', 'knowing', 'believing' as internal and invisible operations of the mind but he claimed that they were as visible as jumping and skipping. How did he arrive at such a conclusion? Let us think of a lump of sugar. There is a lot we can see about it: its shape, colour and structure, and we can estimate its volume and weight. But it has other properties that are invisible which can be made visible. If the sugar is placed in water it dissolves because it has a tendency or disposition to be soluble. We can think of the mind in a similar way: knowing and thinking are its dispositions. How do we make them visible? We test by asking questions and evaluate by the answers. Knowing is not the operation of a ghost in the machine but the exercise of an ability or capacity.[14] But it may be objected that Ryle does not indicate what the source of the dispositions is and often there is a lot more going on in the mind than revealed by its capacities.

Another theory holds that mental and bodily states are identical. Thoughts result from the patterns of the firing of a network of neurons. But it may be objected that if an alternative theory is put forward it also must be the result of physical processes:

> In that case how can one set of motions in the brain be said to constitute the truth, whereas others would constitute a falsehood. If any metaphysical theory is just a physical occurrence in somebody's head, why should one of these occurrences be taken seriously, and other discounted as wrong?[15]

If philosophy cannot supply a satisfactory answer perhaps science can. In 1996, 800 scientists and philosophers gathered for the second biennial Tucson conference, 'Towards a Science of Consciousness', to find an explanation of how processes in the brain create conscious awareness. Two points of view were represented. One was that consciousness is unique, private, subjective, peculiar to the individual and cannot be directly observed by a third person. We have experiences through our senses but what unites all of these states is mind which is the consciousness of being something in them.[16] Knowing how the brain works will be helpful but it does not explain the feeling that 'I am me'. Consciousness could be an irreducible property like time and space.

The other view was that the brain is an information processing machine engaged in competitive activity and mental states are those that win the competition and gain control of behaviour. If the question is raised: 'what are we?' the answer is: we are the organisation of this activity and there is no need to think of some special chamber in the brain that we might label the mind. We are a complex, dynamical system doing its thing! Defenders of the view concentrate on how consciousness is assembled from different neuronal processes and how brain injuries can tear the unity apart. Thus, though we can still see things there is little recognition and patients who have been paralysed claim that their useless limbs still work.

The basic goal in all this, according to Francis Crick, is to find a neural correlate of consciousness. In the 1980s he poked fun at computer lovers and psychologists because they produced models of the brain which bore little resemblance to the real thing.[17] To mention consciousness to them was to show signs of senility! The brain was a complicated 'black box' better left unopened and they were doubtful that the study of neurons would yield results. Crick studied the various regions of the brain in animals which were killed afterwards but couldn't do this with human volunteers! More is known about the brain of the monkey than the human which has so many systems and sub-systems that it remains a mystery. But Crick and his colleagues reported at the conference that changing neural activity can be linked to an awareness of what we see. Illusions show that the stimulus received from an object remains unchanged while what we perceive changes. There is the vase which we just see but after a while two faces are visible on it. Are there then neurons that are specifically related to the act of perceiving? Experiments with monkeys indicated that they might be located at the top of the visual cortex, the higher level of the brain: representation. These link up to the memory neurons. What matters is not the detailed functioning of individual neurons but the pattern of neural activity or organisation. A neuron in the brain is not conscious but needs to come together with others into some kind of organisation or pattern. The link with the mind may be with these patterns of nerve impulses since its activity arises from them. Thus even on this mechanical basis of the brain the emergence of such an organisation cannot be explained by the lower levels.[18]

Roger Penrose argued that consciousness arises from quantum-mechanical processes taking place within tubes of protein inside nerve cells. These micro-tubules are protein molecules which form the internal skeleton of cells and quantum events could multiply there. As we noted, an electron has no location at any particular time until some later event requires it to have one. Until then, it could be anywhere and its position has to be described by a probability function. Penrose suggested that these tubulin molecules may resolve their quantum uncertainties and we would then have an experience but there is no experimental evidence for his theory. However, he seems right in holding that our minds do things that networks of nerve cells and the computers modelled on them can never do. Understanding is a quality that cannot be captured in any form of computation.[19]

What we have seen about the levels of an organism may be helpful. There is a continuity and hierarchy with the mind as an emergent feature and providing a new activity of the whole. It is the organisation of the complex brain with qualities and laws which cannot be deduced from the lower. Thoughts, ideas, love, hate and joy cannot fully be explained by the laws of physical causation which pertain to the brain. In the evolutionary process we see development from lower to higher levels which reveals an organisation of matter to a degree that the highest level of mankind can actually reflect upon the process. In degree of complexity, adjustment to the environment, individualism, creativity and intellect, and self-consciousness, mankind reveals the highest level.

In this way it seems possible to bring the two views expressed at the conference together as complementary. The mind could be a pattern in the brain dealing with information but also the centre of our subjective personal life. It is very complex, as the study of psychology shows, with the unconscious as well as the conscious. The depth of a creative mind is seen in Einstein where his thought experiments produced ideas which were only empirically confirmed later. When we are about to make a decision, neural activity can be observed, but it is impossible to tell what the decision will be. What can be done is to correlate thought processes with observation of the brain and look for patterns. In contrast to the physical causes of the brain operation, the inner life has intentions and motives which require the probing of the why, rather than the how question, unexplainable by mechanical forces.

Thoughts have a power and can produce a feeling of elation or guilt when the brain is not called upon to perform any action. Matter and energy are like two sides of the same coin and the principle of complementarity insists that an object has two aspects. Mind and body interact in the person rather than being two distinct entities. There is an intimate connection between the two for when the body is ill we feel depressed, an intention to do something excites the nerves, the heart-beat accelerates, and muscles move. Anger turns the face red and fear makes the mouth dry and the hands moist. One physical thing can affect another: a brick thrown through the window will break the glass. But if we stand outside the house and wish it to happen the window will remain unbroken. Conversely a brick will not crush the wish: matter cannot destroy thoughts or wishes. I can measure matter but how am I to measure ideas? One belongs to the higher level of my being, the other to the lower. Consideration of the quantum world confirms that causes are more likely to be interpenetrating fields of influence stemming from relationships and interconnected events. There is past influence, present creative selection of possibilities, and the impact of future goals. How far it occurs down the scale of matter is debatable but with us it is clear.

In relating mind and body the idea of correspondence strikes a balance between dualism and materialism. Mind emerges from matter and is not some new entity superimposed upon it but is an activity of the organised complexity. This activity induces a state of correspondence which is not the usual causal one of a link but a correlation.[20] It does not tell us how the link operates but it may receive some confirmation from Penrose's suggestion of quantum effects in the brain where causes are difficult to obtain.

THE PERSON

The word 'person' originates from the Latin term for a mask worn by an actor in classical drama: the actor was acting a role, hence 'person' came to mean one who plays a role in life. We can hide behind the role we play or become so identified with it that the real person never emerges. Sartre complained that it made us 'thing-like', allowing nature and culture to determine what we were: his famous example of the waiter who thinks that the role is what he is. John Locke defined the person as 'a thinking intelligent being that has reason and reflection and can consider itself as itself, the same thinking thing, in different times and places'.[21] Rationality and self-consciousness are singled out but Locke's definition leaves out feeling and willing.

Philosophers distinguish between the biological development of a human being and the morally significant concept of person. Being a person is a moral concept which means a concern for others, not the 'survival of the fittest'. It sets us apart from the sub-human world. It is in the moral community of parents, teachers, friends, employers and so on, that we develop as persons.

As a person we are aware of being the subject or unity of both the mental and physical. Here a distinction can be made between the conceptual and the empirical. The neurosurgeon looking at the brain is following the empirical approach but the conceptual refers to ideas or concepts which involves the meaning and use of language. Consider the question: are all brothers male? The answer is obtained not by observation but knowing that male is included in being a brother. Or the question: has it ever happened that a man has married his widow's sister? The answer lies in the meaning of the word 'widow' and a man has to be dead to leave one. Both examples are conceptual and have nothing to do with the empirical. The point is that in gaining knowledge of persons the empirical and conceptual are combined. It happens also with knowledge of objects. The scientist's experiment is theory laden and guided by what he wants to find. Karl Popper relates that he started a lecture with the words: 'Now I want you all to observe', and waited while the students stared at him and the things in the room. Then one student said: 'What will we observe?' It was a way to show them that observation is governed by selection, interest and the views we already have of the world.

Concepts are abstract but we cannot do without them: citizenship, or names of the days of the week, or nationality. Person falls into this category and arises because we need a term that indicates unity or wholeness. When we explain personal relationships we do not use the terminology of electrons, protons, atoms and molecules but talk about emotions, intellect and volitions. The two aspects do not belong to the same class or category. We think of the first as things and the second as states of the person but as we are made up of atoms we need a physical and a psychological explanation of the one person. We find psycho-physical laws governing the first but with the second, beliefs, intentions and aims, it is so complex that it would be impossible for laws to cover them all.

The concept of person connects with holism. To understand a person in terms of his parts makes a category mistake. Gilbert Ryle writes of a foreigner visiting the Oxford colleges who asks the question: where is the university? It is explained to him that the university is not another institution but the way they are organised and coordinated. The mistake is to think that the university and colleges can be placed in the same category. The parts cannot be confused with the whole and the real cannot be reduced to the observable.[22] The university is not apart from the colleges and the mind is not apart from the body: they affect and interact with one another. In being persons we have reached the highest levels of our being and in this connection we operate downwards controlling the activities of the lower parts. It was the high level activity of mind that ensured survival against animals that were stronger and faster than we were.

We might think of the person as the unified self. From that centre we distinguish ourselves from others by age, sex, height, likes and dislikes and so on. There is a self-awareness and reflection on the impression that we make on others. While animals have consciousness or sensations of cold, heat, hunger, thirst, pleasure and pain, the human self-consciousness is much higher. No other animal has the same degree of self-esteem or an ideal self or a transcendent self which is what we feel we ought to be. The greater the gap between self-image and self-ideal the lower our self-esteem. There is this centre or unifying self around which all revolves but some philosophers and Buddhism deny it and think that we are a bundle of ideas or impressions, not a distinct entity. However, such a stream of impressions seems to apply more to early childhood than later when they become integrated and discriminated so that the distinction between us and others becomes clear. Both Jung and Adler regarded the self as part of the psyche (mind, soul, spirit) which directed aspects of personality. The self develops in interaction with others and self-disclosure occurs in a greater or lesser degree depending on how intimate we are with people and how much we trust them. Such self-disclosure is essential if we are to get to know a person. Relationships are very important in the development of the person and the roles that we play in family, work, as citizens and friends, and express the dynamic activity of the self and contribute to making us what we are.

In self-awareness we realise that we are both the subject (the thinker) and object (what is being thought about): not only the responder to sensations but the interpreter and organiser of such an experience. There is freedom to choose or reject a particular course of action in the process of self-actualisation. There is self-transcendence in that the knowing self transcends the other aspects of self though fully aware of them and itself as the knower. Kant thought that the choices of the transcendental self are determined by reason. We can perceive ourselves as part of nature and bound by causal laws but morally there was the kingdom of ends outside the empirical order and responding to reason alone. Duties, rights and values were moral imperatives by which we judged ourselves, others and nature. When we oppose our desires and feelings we transcend them for some duty that we think we ought to do and if we do not we feel ashamed. We think of someone who does good as affectionate and kind but Kant thought that this stemmed from their nature and not morality. It was the person who while possessing a cold and unsympathetic nature did a kind act who was morally praiseworthy. She acted on the basis of reason which recognised that persons must be treated as ends not means. In the religious traditions such morality requires the transformation of the self, as we will see.

The self we experience remains the same throughout our lives ensured by memory and ability to think about the future. Change takes place but within the same self. While all can reach some form of self-actualisation, which is a life-long process, only a few attain the heights. In the opinion of Jung, the self is the central archetype which unites the personality and only Jesus and the Buddha attained complete self-realisation.[23] The atoms that make up our bodies continually change over the years but the arrangement of them remains the same. The continuity of identity is not in the atoms themselves which the physicist studies but in the pattern.

VALUES

People are concerned today about the loss of the traditional values and blame the media, parents and the schools. Scientific technology advances but moral progress has not kept pace as is seen by increasing violence, break up of family life, attacks on children, terrorism, crime and so on. In the riots which occurred in Los Angeles (1965), 4000 people were arrested, 1200 injured, and 35 killed. Reflecting sadly on the event, the McCune Commission report stated: 'What shall it avail our nation if we can send a man to the moon but cannot cure the sickness of our cities?'

Values arise from our view of what morality is. Morality has been defined as living in accord with moral principles, obligations and ideals, which guide us in our concern for others. Aristotle saw the primary values as justice, generosity, truthfulness, kindness, love, faithfulness and compassion, and scientists need to watch that current genetic work accords with them. In detecting and correcting defects there are benefits but not if there is stigmatisation in being unable to meet a genetic norm.[24]

Where do values originate? Freud pointed to three elements of the personality: the unconscious id which drove the libido and death instinct, the consciousness or ego which modified the id demands, and the superego: the sense of morality. The superego consisted of a conscience which informed us of what was right and an ideal or model derived from parents. But his case studies were limited and the Oedipus and Electra complexes are unacceptable. It is not possible to prove or disprove the structures that he identifies but his work related often to the neurotic. Of course we accept that children do imitate their parents and internalise their values, and conscience induces feelings of guilt, shame and remorse. The superego is learned and points to ideals which change during life as new experience alter our perceptions.[25]

Jung took a more holistic view of the personality than Freud. It was not simply a combination of id, ego and superego but a development of the whole. He wrote about the collective unconsciousness genetically transmitted from common ancestors. It contains images of the earth mother, the fairy godmother, the evil brethren, the wise old man, the valiant hero, god, the trickster and the demon. Alfred Adler set aside Freud's sexual motivation and stressed the child's desire for power which, if he does not achieve, results in an inferiority complex. Today, psychologists pay more attention to the cognitive and social aspects of moral development than did these pioneers of the discipline.[26]

Kohlberg distinguished six stages of moral development. He discovered that at the various stages there were different reactions to perplexing situations and though generally a society required obedience to its rules compassion should be extended when they were broken.[27] Are values relative, then, to certain situations? There is a diversity in the morality of societies. Some have killed infants and invalids and have shown a variety of views about property, women and the honouring of parents. But anthropologists detect a certain uniformity so that most societies prohibit theft and killing, they respect contracts, care for the sick, and so on. Morality is a concern for others. Religion insists that its rules are God-given and its rituals strengthen morality, tribal solidarity, and encourage a spirit of forgiveness. Some societies however do not connect religion with morality but with taboo. Moral rules are utilitarian and can be reinforced or disregarded according to changing needs. Anthropologists seek a balance between rule systems and situational ethics. The problem of the first is that rules can be inflexible and not allow for hard cases but they are based on years of experience and respected authorities, while the difficulties of the second arise as to when we can set aside or 'bend' the rules.[28] Tolerance of the values of other societies is seen as a mark of civilisation. It does not mean relativism since, while we accept pluralism in morality, we continue to hold our own view. There will always be people who live up to the standards of their communities and others who will not live down to them, that is, religious founders, humanists and so on. Morality cannot be reduced to what is socially approved.

Various ethical philosophies try to establish rules for action. Utilitarianism teaches that we act on the basis of the greatest happiness of the greatest number. Kant said that making a man happy is not making him good and it bases morality on the wrong motive: happiness. It neglects the motive or intention of an agent, concentrates on the result, and could consent to the death of an innocent person on the basis of benefit to the majority. He enunciated imperatives so that before stealing I would ask: what if everybody did this? Kant thought that such moral imperatives where prior to experience: a priori, and based on reason. Desires arise from experience: a posteriori, but we need a rational imperative for we are rational. It makes us members of a moral community where things are reasoned and negotiated, and equality ensured. Such a community is an ideal one, however, and being based on reason does not allow for the strength of the passions, self-interest and sympathy. In Kant, duty figures more prominently than happiness, but it is a prime value and religious communities speak of joy here and the bliss of nirvana. He postulated the good will which was based on reason and corrected our values, it judges between what we want to do and what we ought to do. We have the imperative to treat others as ends and not as means. It is rational to respect people. But the more we stress our continuity with the animals and a nature which is amoral the more likely such imperatives are to be ignored. We cannot transfer animal instincts directly to ourselves for we have the reasoned will and, as J. S. Mill pointed out, most of the good qualities that we possess are not due to instincts but victory over them. The vices are more natural than the virtues and conformity to nature has no connection whatever with right and wrong.[29]

We know it is wrong to betray a trust or perform an ungrateful act or tell untruths even though it may mean suffering for us. Some things are intrinsically wrong and we know it intuitively. A philosophy may be reasonable but unjust if it is not concerned for others. Religion postulates rules or principles: 'love and do what you will' (Augustine); 'obey the ten commandments' (Judaism); the eight fold path (Buddhism); the pillars of Islam, but interpreters have seen the need for the exercise of flexibility. Conduct based on the rules is important but it is the inward state from which evil springs that is the problem. Kant was aware of this and pointed inwardly to the good will which overcame the defects. He contended that the other virtues of courage, wit and judgement paled into insignificance compared with good will because they can be misused. Good will reflects character that knows how to use the other virtues and in so doing shines like a jewel by its own light, as something which has value in itself. But Kant separated the subjective and objective – reason and value – instead of making them complementary. We have said that values are not based on facts but a factual situation requires evaluation: it was the facts of disease, decay and death, observed by the Buddha that made him reject his life of comfort and pleasure and seek the highest value: nirvana.

Principles and obligations figure largely in ethical theory but ideals or role models have more power to inspire. Young people have their idols and the educator has to guide them here. The neo-Aristotelian philosopher contends that if a person acquires the virtues of honesty, generosity, justice and so on, he or she will know what to do in a critical situation. A distinction is made between *having* something, characteristics, and *being* something, character. The virtues are an expression of character formed by the habit of repeated choices. If it is asked: why should I follow good values because they do not reward in a selfish society, it can be maintained that virtue has its own reward and results in true happiness: eudaemonia. Of course we do act out of character at times but its development from childhood is essential if we are to realise the values that have welded societies together and inspired them to reach for higher ones.[30]

FREE-WILL

Character is the product of heredity and experience but also conscious decision which involves the will. Animals are determined by their instincts but we make choices and are responsible for them. But if the laws of matter are deterministic and they apply to us where is my freedom? Laws describe what will happen but do not interfere with my choices. The law of gravity means that if I jump off a cliff I will fall to the ground but it does not make me choose to do so! However, it is asserted that when what appears to be a free choice is made, activity in the brain has occurred beforehand and this determines our actions. One answer is indeterminacy in quantum mechanics which reflects the uncaused but it does not solve the problem because whether events are caused or uncaused we are still not responsible for them. Hume thought that necessities do not inhere in the world but only in our way of describing it and there are no necessary connections over time. What the principle of indeterminacy shows is that what happens is not always determined by laws based on some previous state and that initial conditions make prediction difficult. In the quantum context, cause is more like a field of influence arising from a relationship and inner connection of events. If so, many influences will operate on me such as reason and will, resulting in choice. There is much more flexibility than in the old idea of rigid cause and effect.

Psychology is divided regarding free-will. The behaviourist psychologist insists that we are determined by environmental forces which he calls reinforcements: a collection of learned responses to stimuli. The humanistic-existential group asserts that we are unique, free, rational and self-determining. It is free-will and self-actualisation that makes us distinct from animals. We need to take responsibility for our choices, develop our potential and discover the whole self. The biogenic school of psychology argues that we are determined by genetic, physiological and neurobiological factors. Disorder requires physical treatments: chemotherapy, electro-convulsive therapy and psychosurgery. Finally, the cognitive group thinks that we are information processors, selecting information, coding and storing it, and retrieving when necessary. Memory, perception and language are central and we need to use them to control behaviour. Signs of disorder are irrational beliefs about self or others and inability to control them. Since the school goes back to Piaget there is a recognition of the different stages of cognitive development and the remedy is controlling the mind by therapy, Zen meditation and getting rid of unrealistic ideas.

The work of the psychologists and their differences shows the complexity of the human person and that any act of decision stems from the unified self. Influences from the past, thoughts of the consequences, and the demands of the present will all be there. No one factor can be isolated as the reason: some antecedent cause. The decision may be novel and could not be predicated on the basis of past knowledge of the person concerned. Choices are linked with the values that are held but even their powerful influence cannot prevent the exercise of freedom. The problem is understandable when we remember what the brain is like. It contains billions of particles which makes its actions difficult to predict even if we knew the initial conditions. Its sensitivity means that a small change in the initial state can make a very large difference to subsequent behaviour: 'although we know the fundamental equations that govern the brain, we are quite unable to use them to predict human behaviour.'[31]

After years of philosophical and theological discussion the determinism/free-will issue has not been resolved but then it is unlikely either that the paradoxes of physics will be solved: light as a wave/particle or that two events should not be simultaneous for a moving observer and one at rest. We have to accept these paradoxes as complementary, not contradictory. Reductionists say with confidence that we are nothing but machines and therefore determined but how can they be so certain? Was it not Einstein who once said about mathematics which is often regarded as certain: 'as far as the propositions of mathematics relate to reality they are not certain, and as far as they are certain they do not relate to reality'?[32]

SCIENTIFIC SALVATION

The religions offer salvation by the grace of God but science also believes that it can save mankind. Salvation for Richard Dawkins would be liberation from the superstition and the miseries that religion has plagued us with. We can produce the perfect being by perfecting his genetic inheritance. Genetic engineering figures in the media with the call to design the perfect baby and enable it to get a certificate of perfection. Genetic credentials will be the passport to the good life, for in the future social and political judgement will be made on this basis. Nature can be changed not by any mystical grace of God but by sorting out and eliminating faulty genes. We know that environment can shape our futures so parents send their children to Eton, but it is the genetic background that is the most important. We thus have a determinism caused by our genes where autonomy and freedom are ruled out, but while accepting that science can do much to change us, the religious traditions are surely right in contending for the need of a power outside ourselves and the importance of free-will.

WORSHIP/MEDITATION

The religious traditions maintain that worship or meditation is the way to communicate with God. Man is the only animal that engages in such practices. But is it just an instinct even if peculiar to the human animal? If it could be shown that worship was confined to the behaviour of a few tribes then it might be said that it flowed from their culture. But if that is not possible can it be asserted that such an instinct is explained by the evolutionary theory: the survival of the fittest? Unlikely, for:

> Any primitive animal unfortunate enough to suffer a mistake in its DNA copying such that the scrambled-up code gave rise to the message, 'Love your enemies', would have been promptly eliminated by those enemies! No, an instinctive message of that kind could not have been perpetuated in the jungle-like conditions surrounding the emergence of man. The whole tenor of religious awareness and the behaviour to which it leads runs counter to the evolutionary drive towards self-survival, and so cannot be regarded as one of the surviving inherited instincts.[33]

In conclusion we can say that the evolutionary process shows a development from lower to higher levels which reveals an organisation of matter to a degree that the highest level of mankind can actually reflect upon the process. In degree of complexity, adjustment to the environment, individualism, creativity and intellect, and self-consciousness, mankind reveals this level.[34] We cannot agree then with the reductionist that we are nothing but survival machines. We have recognised the value of the method of analysing the parts at the lower levels but to explain the whole by them is too narrow and incomplete. Weizenbaum uses the analogy of the drunken man who is on his knees searching for something under a lamp-post. He tells a policeman that he is looking for his keys which he says he lost 'over there', pointing out into the darkness. The policeman asks him, 'Why, if you lost the keys over there, are you looking for them under the street light?' The drunk answers, 'Because the light is so much better here.'[35] The reductionist refuses to move from the circle of light cast by the assumptions of physical science to the darkness beyond but the light must shine on the whole man: his emotions, values, thoughts, and choices and these require a personal explanation.

In the next two chapters we consider what insights regarding the world, mankind and God, can be gained from the religious traditions.

7 The Indian Tradition

Elizabeth Ann
Said to her Nan
'Please will you tell me how God began?'
And Nurse said, 'Well!'
And Ann said, 'Well?
I know you know, and I wish you'd tell.'
And Nurse took pins from her mouth, and said,
'Now then darling – it's time for bed.'

Elizabeth Ann
Had a wonderful plan:
She would run round the world till she found a man
Who knew exactly how God began.

A. A. Milne

Elizabeth Ann is raising a profound question and no one would find fault with nanny for not being able to answer! But we can enquire into what God may be like and see how various religions picture him. We have selected six religions from the Indian and Semitic faiths and will concentrate on matters relevant to their view of the world, mankind and God. In this chapter we consider Hinduism, Buddhism and Sikhism, and in Chapter 8, Judaism, Christianity and Islam.

HINDUISM

The Hindu view of the world is nearer to the scientific than the Semitic since it sees our planet as one of a vast number of worlds which alternate between a state of evolution, repose, and ultimate collapse when a new age begins. In this connection the model of God as Shiva appears, engaging in a self-expressing dance of everlastingly creating and recreating the universe.[1] It is not in competition with the scientific account, for it is mythical with various heavens and hells and a deification of nature. Time moves in a circle in accord with the seasons and human birth, death and reincarnation. Little difference is made between the mental and material except for the soul or self or principle of life (atman).

Hinduism is a vast religious phenomenon without Founder, creeds, church organisation or central revelation, but it has priests, temples, rituals and many gods. There is a great diversity of belief, practice, and gods, arising from tradition, social needs and the Indian scriptures or veda composed between 1500 and 800 BC. They consist of collections of hymns, priestly commentaries (Brahmanas) and philosophical treatises: the Upanishads. Village Hinduism has many gods but behind all and within all is Bhagvan, the impersonal supreme Spirit, or Brahman (neuter), the Absolute or Ultimate reality.

Brahman means sacred power or energy, invoking the sense of awe. It is not an object in the world but the source of it and can only be described negatively: 'not this; not that'.[2] But it is generally accepted that it has the attributes of being (sat), consciousness (cit), and bliss (ananda). Union with it requires an ascent to a higher level of consciousness and knowledge of a particular kind: jnana or insight, obtained by meditation. But this concept of Brahman was too impersonal, hence the development of the more personal gods such as Brahma, Vishnu and Shiva. Vishnu is worshipped in the forms of Rama or Krishna and they appear in the two great Indian epics, the Ramayana and Mahabharata. As avataras (descent) they are the intermediaries between Brahman and mankind. Their lives may have some historical core but it is the meaning that is intended: to picture the human face of deity.

THE WORLD AS THE BODY OF GOD

The universe emanates or evolves from Brahman who is behind all things. One image is that of Krishna, the incarnation of Vishnu, casting his seed in 'Great Brahma', that is, material nature. Thus from the primal sexual act the universe comes into being (Gita. 14.3). It is a mother image of deity but the process can also be compared to 'bubbles in water' as the worlds arise from, exist in, and dissolve into the supreme Lord, 'who is the material cause and supporter of everything'

(Atmabodha 8). Emanation from Brahman is a closer relation than that of a creator distinct from his creation and the union is strengthened by making the world the body of deity. In classical Hinduism, three philosophers discuss the issue: Sankara, Ramanuja and Madhva. Sankara lived from 788 to 820 AD and taught that Brahman and the soul were one: 'That is the true; this is the self; that art thou.' (Chandogya Upanishad VII. 13). The world was an illusion (maya) emanating from Brahman who exists on two levels: the higher and impersonal and the lower personal level. It is the lower that is experienced in the manifestation of the gods. Brahman uses illusion (maya) in a magical way concealing itself so that we think its manifestations are real but they are not because of the unreality of the world. How could a world of suffering and death be real?

Knowledge is insight (jnana), attained by meditation, into the higher level of Brahman and it is superior to the worship of Isvara at the lower level. Brahman is perfect and unchanging at the higher but at the lower contains potentialities which realise themselves in the world by emanation or procession. But it may be objected that if Brahman at the higher level has consciousness does it not imply the personal quality of mind? What does Sankara mean when he says the world is an illusion? Illusions such as mirages occur in the world and delude us but we are able to recognise their illusory nature. I can mistake a rope for a serpent but the rope is there and I can distinguish. He quotes the scripture:

Lead me from the unreal to the real;
Lead me from darkness into light;
Lead me from the mortal to the immortal.
(Brhadaranyaka Upan. 1.3.28)

Think truly, this life is but a dream.
With mind fixed on truth one becomes free from attachment;
To one freed from attachment, there is no delusion;
Undeluded, the soul springs to clear light, free from all bondage.

The dream is real but when we awake it vanishes. The world is like a dream but when we view it from the mystic higher level we understand that it is illusory. Hence the advice: 'Give up this Maya made world, gain true knowledge. And enter the path to Brahman.'

It seems to mean that we are deluded by the material world and if so then he is in accord with other religions. But if the manifestations of Brahman are illusory then they do not convey real knowledge of the world so we have an unknowable God. The stress on insight into reality, however, is valuable and though gained from a different method, that is meditation, has a parallel in science: seeing into the reality of things. Sankara identified the world with Brahman. Some dispute it but admit that their interpretation is 'a very stretched one'.[3] He based his view on the Upanishads but ignores those passages that distinguish Brahman and us. Some of the Upanishads – Svetasvatara, Isa and Katha – are more theistic (belief in God who is distinct from the world) than monistic. It is in the early Upanishads that monism is prominent,[4] but the distinction is stressed in the favourite scripture of many Hindus, the Bhagavad-Gita or 'the Song of the Lord', probably composed in the fourth and third centuries BC. Sankara's distinction between higher and lower levels is useful with regard to scripture since we need to move from the lower literal level to the higher symbolic level. But it is incorrect to make meditation higher than worship and if the soul and the world are identified with Brahman where is their autonomy? It would also restrict the 'otherness' of the deity, and, if worshipper and the object of worship are one, how can there be any worship? We would be worshipping ourselves.

Ramanuja was not satisfied with Sankara's position. He lived in the 12th century and encouraged the worship and devotion (bhakti) of Brahman whom he considered to have personality. Both the world and souls are real and while the individual soul shares in the essence of Brahman it is never wholly one with it. The world is the body of Brahman who is in it as the soul is in the body except that the deity is totally in control of the world and transcends it. The view is identity-in-difference. The world emanated from Brahman without the use of material outside himself – the neuter gender is set aside since Brahman is the highest Self – but there was no beginning for the universe is continually being created. Ramanuja denied that Brahman was pure consciousness for this does not agree with Vishnu's personal revelation in Krishna.[5] He expounds the Vishnu

Purana (epic poems, six each for Vishnu, Shiva and Brahma) rather than the Upanishads but finds his view confirmed in the Isa-Upanishad where 'he' is used instead of 'it': 'He is the seer and thinker, pervading all, self-existent' (verse 8). This more personal view of deity is combined with his being self-existent: not being dependent on anything else. There is also the Gita where the deity is characterised by thought, bliss, wisdom, strength, lordship and grace (13.9). The Svetasvatara has a triad consisting of the Lord, the material world and the individual soul, and the distinction of each is retained. But at the same time Brahman contains the two elements of the world, matter and soul, within himself. The womb of the world is Brahman, according to the Gita, and it contains matter potentially (prakrti).[6] Ramanuja contends that the emanation of the world from Brahman is not of necessity but occurs from free choice for there is no external force constraining him to bring it into existence.

If Sankara stressed non-dualism (advaita) and Ramanuja qualified dualism (visishtadvaita), Madhva who lived in the 13th century AD, embraced dualism (dvaita). He opposed Sankara's interpretation of the statement of the Upanishads: 'That self; that art thou' (sa atma tat tvam asi). Since classical Sanskrit is written continuously and not broken up into separate words a possible translation could be the converse of 'that art thou' namely 'not-that art thou'![7] According to Madhva, Brahman, souls and matter are eternally distinct and salvation is drawing close to the deity, not being united with him. The same stress on the grace of God emerges as in Ramanuja but it is bestowed on those who live righteous lives: deeds (karma). That evil deeds will merit eternal damnation – separation from Brahman – accords with the Christianity prevalent in the area where Madhva lived – but its permanence is not the usual Indian view.

Of the three thinkers the theism of Ramanuja was the most far reaching for its bhakti impressed Kabir (1440–1518) who in turn influenced Guru Nanak, the founder of the Sikh religion. The three Indian philosophers based their view of God and mankind mainly on the Upanishads but the Svetasvatar Upanishad (c. 300 BC) confirms Brahman as a personal being who can be worshipped and is called Bhagavan. Vishnu was regarded as one of the manifestations of Brahman and in the

Gita he becomes the supreme God. He is a kindly deity, who appears in the form of Krishna to Arjuna as he makes ready to fight a battle and tells him that he pervades the universe and that the soul is indestructible. Arjuna addresses Krishna and is granted a vision of the amazing numinous power of Vishnu as imperishable and changeless Being and the One who can grant liberation by his grace from the round of samsara (rebirth):

> On Me thy mind, for Me thy loving service (bhakti),
> For Me thy sacrifice, and to Me be thy prostrations:
> Let (thine own) self be integrated, and then
> Shalt thou come to Me, thy striving bent on Me.

There is also the love of Vishnu:

> In whatsoever way (devoted) men approach Me, in that same way do I return their love. Whatever their occupation and wherever they may be, men follow in my footsteps. (4.11)

At times it is difficult to know from the Gita whether souls merge in the deity or live separately but in close union. But it seems clear that to know Vishnu is by bhakti or loving devotion, and the personal is given priority over the impersonal as Vishnu the Supreme Person displaces Brahman.

HINDU MODELS OF GOD

Commentators on Sankara see various levels of deity: impersonal and personal the Lord of bliss, knowledge and unlimited existence without contact with the world; qualified Brahman as causal and effected and the incarnate Vishnu. The first defends its impersonality, the second the bliss which does not depend on anything external, and the third and fourth, his contact with the world. This relegation of the personal to the lower level is unsatisfactory for the personal is the highest that we can conceive.

It emerges in the incarnational model in connection with Vishnu and Shiva. They are avataras (descent), the human

intermediaries between Brahman and mankind and display
weaknesses in their lives. But the record is mainly mythical for
Hinduism is interested in what they symbolise. These avataras
which include the Buddha belong to the concept of cyclic
time and appear when needed: 'Many a birth have I passed
through ... For whenever the law of righteousness withers
away and lawlessness arises, then do I generate Myself (on
earth)' (4.57). Vishnu transcends Brahman itself (10.12;
14.27) and in him nirvana subsists (6.15). To take the record
of Vishnu's incarnation in Krishna literally would be to see
him as a joker indulging in love affairs but the Hindu stresses
the symbolism and sees his life as reflecting the spiritual love
between Vishnu and souls.

Then there is the transformation model, for in bringing the
world into being Brahman transforms itself. We think of milk
turning into curds or the sky changing colour so by analogy
we could think of a change taking place in deity. The effect of
emanation has consequences for the cause.[8] Brahman is both
Being and Becoming when we think in this way of it. Then
there is the Cosmic Self model: 'truly the ruler of all beings,
the king of all beings. As all spokes are held together in the
hub of a chariot wheel, just so, in this Self, all being, all gods,
all worlds, all bodily organs, all these selves are held together'
(Brhadaranyaka Upanishad 11.5.15). The four aphorisms of
the Upanishads are: 'Consciousness is Brahman', 'That Thou
art', 'The Self is Brahman', and 'I am Brahman'. Brahman is
the universal Self and we are part of it. But the part is not to
be equated with our ordinary ego or empirical self, it is the
ideal ego. Thus the Hindu looks inward and by meditation
tries to realise it to attain union with Brahman. In our discussion of the human self we saw that we cannot observe the self
but are aware of it. There is a sense in which the self transcends time and space and by analogy we can think of
Brahman transcending the world. Religions teach a denying
or dying to the ordinary self in order to achieve union with
God. But if I am identical with the Cosmic Self so are you: the
belief that we are separate selves is an illusion. The interdependence that we noted in the quantum world has a similarity with this.

The world of the Hindu is full of divinity because he sees it
in all things: in rivers, mountains, animals and so on. Notable

people such as Gandhi have a high degree of divinity reflected in Einstein's statement about him: 'Generations to come, it may be, will scarce believe that such a one as this ever in flesh and blood walked upon the earth.'[9] There are also goddesses and Durga the wife of Shiva is regarded as the personification of divine energy. But the villagers can unify their worship by remembering that Brahman lies behind all things.

KARMA

Like other religions, Hinduism deals with spiritual ignorance from which ensues sin, suffering and alienation from Brahman. Everyone must do what is required of them in their station of life which has been determined by the law of karma. In both Hindu and Buddhist thought karma has various shades of meaning: the moral character of an action, the deeds which determine destiny in a future reincarnation, the causal laws that we reap what we sow, and the force governing the destiny of mankind. Karma enforces duty: Arjuna must fight for he is a warrior and is reassured by Krishna that he cannot kill the immortal soul and thus should not worry about the death of his relatives. The villager wants to survive, acquire merit and salvation, avoid sin (pap) and know that pollution is to be avoided for the worship (puja) of a god requires purity. He will follow all the rituals and perform the required acts to fulfil his karma. But the overall impression given by Hinduism is that while sin or evil must be avoided there is more stress on ignorance, the need to get the right insight into life, and union with Brahman. The alienation from God which appears in other religions, apart from Madhva's teaching, is not prominent, for divinity is present in all.

The law of karma operates as the expression of the will of Brahman and implies its power and sovereignty. It is more extensive, not only placing the Hindus in their societies and caste, but also governing their rebirth. When it is connected with grace then there is the action of God which modifies the view that Hinduism differs from the Semitic religions in not having the model of God as agent and continually seeking Brahman in inner experience. What appears to be the difference is that such a model is primary in the Semitic. Like all religions Hinduism has a developing concept of God. The personal rather than the impersonal gains prominence in the devotionalism of South Indian religion and with outstanding people like Gandhi (1869–1949).

Hinduism is tolerant of other religions recognising that there is the same God: 'He is one, [though] wise men call him by many names' (Rigveda, 21,164,46). Just as a winding river at last merges into the sea, all paths lead to Brahman. Thus the philosophers worked out a synthesis of the various religions' concepts of God as fragmentary revelations of the one God, an idea which in more recent times has surfaced in Christian theology.[10]

BUDDHISM

The Buddha means the Enlightened One, assumed by Gautama Siddhartha, after his experience of enlightenment (531 BC). The young Siddhartha reared in luxury was dismayed by old age, illness and death which showed the impermanence of life and the centrality of suffering. The term 'dukkha' means sorrow but in the first noble truth enunciated by the Buddha it includes imperfection, impermanence, emptiness and insubstantiality. The soul or self shares this impermanence and is rejected as a central entity. Meditation is the way to enlightenment. After his death the Buddha's followers formed a monastic order: the Sangha. The religion divided into Theravada (the doctrine of the Elders) which is active in Sri Lanka, Burma and Thailand, and Mahayana (the great vehicle) located in the East and North of Asia. The scriptures are described as three baskets (tripitaka) with narratives concerning the Sangha, discourses of the Buddha and seven books of doctrine.

Before he died the Buddha said: 'After me your Master will be the Doctrine itself', but his disciples gave him the status of an infallible Master and Model. Each follower was to prove the truth of his teaching in his own experience: 'Be, each one of you, your own island, your own refuge; do not seek another refuge. It is in this way that you will reach the high place of the Immortal.'[11] The truth of the faith must be tested by experience rather than disputing texts of scripture, for words can only point to the highest reality. The Buddha did not think that the gods could help anyone to find nirvana, that is, the highest good, since they were in need of it themselves. Man's body is worthless, decaying and dying, there is no soul or permanent ego, but a collection of acts or an aggregate of mental and material elements (skhandas) which are compared to foam, bubbles and mirage. It is a state of change or unreality. The skhandas are sensations, ideas, predispositions, tendencies and thoughts received from the world and consciousness arises out of them. There is no centre or 'I' which is an illusion: we are a bundle of experiences. But what then passes over in samsara (rebirth)? The answer is karma or deeds,[12] but some Buddhists give prominence to consciousness as a stream flowing uninterruptedly into another body.

The Buddhist is like the reductionist in science for he reduces the human into smaller and smaller units so that we have no longer any extension or duration, resulting in the conclusion: 'there is no agent, one finds only action. A way exists, but there is no wayfarer'.[13] We have seen that energy and matter are like the two sides of one coin and the Buddhist maintains that the organisation of energy at one stage influences what the flow will become at a later stage and it extends beyond death. The soul becomes the process of energy which animates another body in the cycle of rebirth. It is like lighting one candle from another or, as the Buddha said, the movement of the leech, that is, before we leave this life we have a grasp of another.

Like Sankara who was influenced by Buddhism, the world is seen as a mirage which deceives us into thinking it is permanent and can fulfil our desires. But all is impermanent and in such a world how can there be a permanent soul or self which claims: 'this belongs to me'? From the deception springs the desire for power, possessions and so on, resulting in suffering. What we need to know are the four noble truths: all life is suffering, it is due to desire, desire must be eliminated and this can be achieved by following the eightfold path. It means a right view of the insubstantial nature of objects so that they do not arouse desire, the relinquishing of them by becoming a monk, the rejection of false aspirations which lead to greed and hatred, and right living. Any livelihood which harms others, the sale of alcohol or trafficking in women and slaves must be avoided. Right thinking and meditation rounds off the path which leads to nirvana. It is a state of detachment from all desires and thoughts, 'neither awareness nor absence of awareness'. The use of reason is a stage towards the goal but eventually the mind is emptied of thought to grasp the inner nature of things lying behind ordinary experience. Thus the Buddha spoke of 'a refuge' or 'island' or 'another shore'.

This was basically the position of Theravada Buddhism but Mahayana emerged and taught that Theravada was too narrow, for, both monk and the ordinary lay person should be able to experience nirvana. Grace or help is needed and it pointed to the Bodhisattvas, persons who have attained final enlightenment (bodhi), but renounce entry into it in order to help others. Differences also arose about the meaning of nirvana: is it negative or positive? The root meaning of the word is 'a blowing out' or 'extinction' so it seems to be a void or nothingness (sunya). But other philosophers saw it as an indescribable state of bliss: 'the summum bonum'. There is nirvana with substrate: enlightenment now, and nirvana without substrate: occurring at the individual's decease. What is needed in Buddhism is knowledge gained by meditation: the root of evil is ignorance.

A number of questions arise. If nirvana is permanent how can I, an impermanent entity, experience it? Can the permanent and impermanent exist? We recall that Plato held that the forms were permanent but Heraclitus stressed the impermanence of everything. If we think of the forms existing in another world, and the Buddhist does think of nirvana as transcendent, both can be held together. We have also seen that order or necessity is permanent in the scientific view of the world but combined with change and chance. Self-identity does appear continuous throughout life preserved by memory, and in the development of Buddhist psychology an attempt was made to contend for the individual (pudgala), but it was condemned at the third Council held at Pataliputta in 247 BC. The Buddha was evasive when he was asked by his disciple, Malunkyaputta, about the relation of soul to body, the nature of the universe, and whether the Buddha would exist after death. He replied that one did not need an answer to such questions before deciding to lead a holy life just as a man, wounded by an arrow, does not say that he will not allow it to be removed before he knows who fired it or what type of arrow it is! But some Buddhists admit that there is a relation between the last thought of this life and the next, indeed it conditions it. Is there then not something special about mind or consciousness which makes it central?

In the Mahayana development, worship and veneration for the Buddha was a natural instinct even though he had warned against it. More changes occurred with the idea of many Buddhas, the introduction of female deities and the concept of many worlds with a Buddha for each. Some thought that all the Buddhas were only different manifestations of a fundamental Buddha for it was unlikely that Buddhas would come continually to a limited world of matter. But the doctrine of the three bodies or three different aspects of the Buddha was developed so that there is the eternal teaching or essence (dharma), the historical Buddha and the transcendental Buddha. A similarity has been seen with the Christian view of the eternal Word, the historical Christ, and the transcendent Lord, but the resemblance is superficial because in the Lotus Sutra scripture there are many Buddhas in heaven.

In Mahayana the Bodhisattvas are regarded as compassionate saviour figures and act as acolytes to the Amida Buddha, the transcendent Buddha of Infinite Light and the personification of infinite mercy, wisdom, love and compassion. Grace flows from Amida but this devotional current is not in accord with the Buddha's teaching nor are the militant and magical-ritual groups active in some countries. As in Hinduism, divinity can be ascribed to any good person and in Mahayana Buddhism once an individual has attained enlightenment, he or she is entitled to be called a god. Buddhists in general are opposed to the concept of God as creator for if such a Being existed he would not have made a world of suffering. The view is held despite the answer which the Indian tradition finds in the law of karma and which is accepted by the Buddhist: we suffer because of our desires and deeds. But the Mahayana brought back the gods, instinctively recognising the need for an Object of worship. The Buddha did not deny the existence of the gods but considered that such a question was unimportant. After his enlightenment he was reluctant to speak about it but Brahma Sahampati, who was recognised as a great king of the gods, pleaded with him to share his experience with him and others. Thus they were responsible for the world hearing of the Buddha's experiences.[14] The Theravada regard the Buddha as infallible and the Mahayana think of him as omniscient. Even in Theravada there were schools, rejected by the majority, that concentrated not on the historical Buddha but on his being a channel of divinity. The real Buddha was transcendental, without imperfections, omniscient, omnipotent, infinite and eternal with the historical Buddha being a magical creation of the transcendental Buddha, a fictitious creature sent by him to appear in the world. In the Mahayana the historical Buddha can fade away and leave the Buddha as the embodiment of Dharma (the Buddha's teaching). The Buddha is a type of manifestation and only one of a series of Buddhas who have appeared on earth throughout the centuries. Thus the Mahayana populate both heaven and earth with Buddhas.

It is also possible to equate nirvana with the concept of God for even in Theravada it is immovable, everlasting, deathless, permanent and so on, which are predicates of the deity in the theistic religions. The state that the mystic reaches is comparable to the Hindu who experiences Brahman. He arrives at the other shore where is the immortal, bliss, safety and the island of refuge. It is easy to see why Mahayana became more popular than Theravada because in the latter the only logical Buddhist is the monk, whereas the salvation offered by Amida is for all and there is hope for the faithful that upon death they will be reborn in his Paradise: the Pure Land of the West which is a place where conditions are right for realising nirvana.

SUNYATA OR EMPTINESS

We have referred to three basic concepts of Buddhism: nonself, nirvana and emptiness or the void (sunya). All are interrelated but what does sunya mean? The short answer is a denial of all conceptual constructions in relation to ultimate reality. We think that emptiness is nothing but here it means indescribable, and it can also be connected with impermanence and insubstantiality. The belief in a permanent self causes competition and strife instead of cooperation. Zen

Buddhism seeks to dispel the illusion of such permanence by insisting that nothing has real independent being but is a temporary event in the process. Event is where a certain characteristic shows itself at a point of time, due to conditions which are themselves similar events. Thus each event is empty in that it simply arises from the conditions which bring it into being. Buddhist's atomism is more perceptive than the Hindu where atoms are everlasting, tiny building blocks of the universe. With the Buddhist the world is a vast set of short processes, the one giving rise to the next according to a complex pattern. It has a similarity with the process philosophy of A. N. Whitehead, which we consider in a later chapter.

The mistake that we make is thinking of objects and ourselves as possessing natures or realities indicated by the words. Zen, which originated in China, assimilated ideas from Mahayana Buddhism and Chinese ideas about the Tao or the great Way. If we put together the statement of Zen: 'no dependence upon words and letters', and that of the Tao: 'the Tao that can be expressed is not the eternal Tao, the name that can be named is not the eternal name' we have the inadequacy of language to express what Ultimate Reality or God is. This is non-realism embraced today by some Christian theologians. Thoughts correspond to nothing outside the mind, hence Nagarjuna called into question all the concepts of Buddhism itself! They are relative and depend on temporary conditions. If we think of a young man who lives in a large city and comes to hate the noise and violence he can react by saying: 'this is bad I must move to the country', which he sees as 'good'. But he soon becomes bored with the silence and lack of activity of the country and returns to the city which has now become 'good'. Values are relative and do not possess any intrinsic nature for emptiness means that nothing has a fixed nature or essence. Zen does not stop there but considers that it explains the openness of things so that we can experience them in various ways.[15] Zen believes that our concepts actually get in the way of such understanding and lead to desiring objects, with consequent suffering.[16] But we normally analyse concepts and try to refashion them in order not to desire objects but to get nearer to the truth of what reality is. This is critical realism which recognises that we do not know reality in itself but tries to get nearer the truth about it.

Nirvana is the only stable reality and everything must be directed to its search. When we consider it in its positive form it is akin to the idea of the Absolute (inner essence of the cosmos in Mahayana Buddhism) among the philosophers and God among the worshippers. Hence when a Buddhist speaks of emptiness he is thinking of a nirvana which cannot be expressed in our speech. Most scholars see it as impersonal but if we say that it is the dispositional state of the person who has achieved it then it is personal. But nirvana may not include individual survival for if we are simply a sequence of states there is no point at death of speaking of survival.

Nirvana in the Theravada is not God or the Absolute, it is not the cause of the world or personal but a transcendent state. By way of contrast, in Mahayana nirvana is identified with the Absolute and the transcendent seems to be some kind of substance. Hence the Buddha is a revelation of a 'hidden' ultimate reality and we have the three body doctrine of the Buddha. In Pure Land Buddhism, nirvana tends to take second place to faith in the Buddha which brings Buddhism close to theism. Devotion and worship are evident and mysticism leads to an experience of God. Christian scholars do not hesitate to see theism in this, for the Buddhists are using concepts which transcend the world and apart from nirvana there is dharma which is the substance of the Buddha's teachings. These concepts, while present in our world, transcend it and are analogous to the term 'God'. Making the equation between God and the Absolute, Hans Kung writes:

> If God is truly the Absolute, then he is all these things in one: nirvana, insofar as he is the goal of the way of salvation; dharma, insofar as he is the law that shapes the cosmos and humanity; emptiness, insofar as he forever eludes all affirmative determinations; and the primal Buddha, insofar as he is the origin of everything that exists.[17]

Nirvana is elusive for to get it we must not be attached to anything even the Buddha, hence the statement: 'if you meet the Buddha kill him!' But such a demand with the self-effort needed to attain nirvana emphasises individualism and how is this related to the Sangha where there could be an attachment? And do we not desire nirvana? The Buddhist recognises the objection but replies that if we desire it we will not get it! Nirvana cannot be attained as long as it is sought and it is only when this last and purest desire is extinguished that the goal is attained.

What is analogous to the scientific picture of the world is the interrelation of things and that emptiness is an openness of systems such as ourselves to receive and react. When we are empty of the illusion of self and its demands we are in a position to receive freedom, wisdom, compassion: nirvana. But if the Buddhist says the self must be emptied, what of the cosmic self? Some Buddhist scholars see in the concept what we noted in Hinduism, a change taking place in God.[18] Masao Abe, for example, is not satisfied with the idea that God is Being and Becoming, since if we say God is love we are asserting more than he has become human: he has emptied himself. He believes that it is the view of the Catholic theologian, Karl Rahner, though he thinks that Rahner is not clearly stating his position. It means the self-abnegation of God whereby he identifies himself with sinful humanity and sacrifices Godself.

Everything, including God, is interdependent: nothing exists by itself because nothing possesses a fixed and enduring selfhood. The interpenetration of the human self and God will mean that the substantial otherness of God is rejected but God can be retained as the Lord of grace. In Buddhism, sunyata being open and boundless, lacks value judgement and direction in history hence it needs to be negated and emptied! Abe then takes an important step: sunyata must be transformed into a personal God as exemplified by the Amida Buddha. Conversely, in order that interpenetration between the self and God takes place and his otherness and transcendence be overcome God must empty himself.

The argument results in the conclusion that there is a self-emptying or transformation in God. We have a model of God which, according to Abe, has an affinity with the emptying of God in the incarnation of Christ (Phil.2). But it appears to be bought with the price of permanence in God for Abe rejects the Vedantic notion of Brahman as the sole and enduring reality underlying the universe because of the impermanence of everything and the interdependence of all. Nothing possesses a fixed and enduring selfhood and the recognition of this is sunyata: the ultimate principle. But our reflection on the scientific view of the world is that necessity and change exist together and when applied to God would mean that he is a permanent Being who yet engages in Becoming. Amida is a personal being but the ultimate principle of sunya understood in connection with nirvana would be beyond 'personalisation'. Later we will think again about 'nothing' with regard to creatio ex nihilo and the scientific discussion of quantum fluctuations in a void. But the thought that God sacrifices or negates himself has a parallel with the Christian belief in God incarnate. In that Christ refused to save himself in order to save others there is some similarity with what the Buddhist is saying though the denial of the self as a permanent entity remains debatable.

SIKHISM

Guru Nanak (1469–1539), the founder of Sikhism, was influenced both by Hindu and Moslem thought and experienced enlightenment when he was about 30 years of age. One day Nanak failed to return home but after three days he reappeared and explained that he had experienced a revelation of being at the court of God, offered a cup of nectar, and told to preach the name of God:

> I was a minstrel out of work;
> The Lord gave me employment.
> The Mighty One instructed me;
> Night and day, sing my praise!
> The Lord did summon this minstrel
> To his High Court ...
>
> On me He bestowed the Nectar in a Cup,
> The Nectar of His True and Holy Name ...[19]

He travelled extensively preaching and performing miracles until his final settlement in the Kartarpur. His followers were known as sikhs or disciples and he was able to choose his successor before he died. The authority of the movement was centred in these various successors or Gurus but eventually it revolved about a holy book, the Guru Granth Sahib or Adi

Granth, that is, first book. The term 'guru' is used for designating a holy man but in Sikh writings it is sometimes a synonym for God, thus the Guru Granth Sahib is the voice of God for the Sikh. It consists largely of a collection of hymns by Guru Nanak and other Gurus and is venerated in every place of worship (gurdwara), as an oracle for guidance in the practical affairs of life.

Nanak had a high view of God seeing him as eternal, creator, timeless, self-existent, and revealing himself by grace. He is present as light in every heart whether recognised or not. Nanak refers to the Name (nam) of God as meaning his nature and urges his followers to repeat it in their devotions to achieve mystic union with God. Nanak accepts the Hindu triad of Brahma, Vishnu and Shiva, but there was no question of God needing them or the cosmos. He created the world for his own delight and by his will he continues to sustain it. Hence there is the idea of continual creating and the pervading of the world. He is a God of grace (nadar) and reveals himself everywhere in the world but we do not grasp it by reason, it is the gift of God. Preparation for grace can be made and in that sense the believer is not passive but active and open to God.

The world is maya (deception), but in Sikhism it means delusion which does not deny the reality of the world but the values which it represents. The world is not evil in itself but to see it as permanent and having lasting values would be falling into the mistake of making the truth into a lie. We must turn from worldly values and seek the revelation of God through his word (sabda) which is uttered by the Guru. It is the voice of God within the heart. Mankind is blind and self-centred (haumai) and such egoism causes violence, sorrow and doubt and the inability to see God. The problem is the ego. The person who hears the voice of God and disobeys (manmukh) is contrasted with the one who obeys, gurmukh. It is the love of self which causes disobedience and it needs to be dealt with. Five sins are mentioned: lust, wrath, avarice, worldly love and pride, which stem from the ego and this must be replaced by union with God. Sikhism is a life-affirming religion applied in a world which is not evil. It denies asceticism, the physical renunciation of the world and the use of yoga to overcome the senses. Nor did Nanak think that ritual and external exercises of devotion lead to salvation, for we need to get beyond them to know God. He questioned pilgrimages, giving of alms, ritual bathing and the use of idols in the home and temples. There is a need for a spiritual religion and a direct relationship with God: 'I would bathe in the holy river if I thought I would gain His love, But without it the ablutions are useless'.[20]

Nanak accepted the Hindu law of karma whereby 'past actions determine our garment', that is, reincarnation. There is a strong emphasis on the sovereignty of God but a liberal interpretation of Sikh belief is God permitting what comes to pass but not ordaining it, leaving room for human autonomy.[21] The unity of God and mankind is so close that it is like the drop in the ocean and the ocean in the drop, cold in the midst of heat, man in woman and woman in man. It looks like being absorbed into God but a better understanding is identity-in-difference. It is the recovery of a unity which exists eternally. In such a union (sahaj), the believer is able to recognise both God and himself: individuality is preserved.

What is the status of of Nanak? There are various views within Sikhism itself but in general it is insisted that he is the saviour, prophet and true enlightener and as perfect as God himself, but not God. In him God has manifested his spirit. The view rules out incarnation, for Nanak is human and not a supernatural being, and Sikhism avoids the use of the Hindu avatar. Yet it is said that he possesses divine attributes being without error, perfect as God Himself, his utterance is the utterance of God, manifests his spirit, is sinless, and a saviour. This is balanced with a stress on his humanity for it is argued that unless he was subject to the same laws as every other human being he could not be an exemplar and effective in saving man. Though Nanak partook of divinity he is not equal to God. What we can say is that his guruship is derived from him. He achieved the status which implies that man can in some sense become divine but to see him as an incarnation of God would be incorrect. However in a Hindu context there would be no problem in seeing him as divine.

The symbols used by Nanak, word (sabad), name (nam), divine preceptor (guru), divine order or will (hukam), truth (sach), and grace (nadar) are all related. Nadar shows that God must reveal himself if he is to be known and sabad means a form of divine self-expression: 'the word is my Guru'. The action of God appears to take place through the sabad which not only implies a model of God as agent but some similarity with the Word (logos) in Christianity.[22] The conclusion would seem acceptable to those Sikhs who speak of his Guruship being derived from God but not to those who designate him as a prophet.

Whether Nanak would have regarded himself in the way he is portrayed in modern Sikhism is debatable but we know little of what he was really like. His life is contained in the janam sakhis which are collections of stories relating to historical events, legends and mythology. Hew McLeod applies a historical and critical approach to them and concludes that little can be known about Nanak's life. Some Sikh scholars agree with him but others disagree and use the stories as a basis for doctrine and ethics. The difficulty in the recovery of the historical person is not peculiar to Sikhism and applies to other religions. Perhaps the best estimation of Nanak is that of a charismatic prophet who has spiritual gifts and brings a revelation. As such he is the human model to follow for he portrays what God is like.

CONCLUSION

In this chapter we have seen some kind of unity about what God is like emerging in the various religions. God is both transcendent and immanent, impersonal and personal, and – with the exception of Buddhism – creates and sustains the world. Creation is due to the will of God. The impersonal nature of God predominates in Sankara: Brahman is only personal on the lower level. But the personal is the highest form that we know and there is in the development of Hinduism a reaction with Ramanuja against such impersonality and in the Gita, Vishnu is the Supreme Person. With Ramanuja the personal God is the Lord, the impersonal is only one aspect, and the pantheism of Sankara is replaced by a form of panentheism. Both aspects are in Sikhism. In Mahayana Buddhism, there emerges the transcendent Buddha of Infinite Light who is the personification of mercy, wisdom, love and compassion. And, nirvana, interpreted positively, could be equated with God.

The gaining of knowledge is essential in all three religions regarding the world, mankind and God. Lack of it is the defect in mankind in Hinduism, desire lies at the root of our suffering in Buddhism, and disobedience in Sikhism. In the three, the ego must be transformed. Insight is important, connected sometimes with meditation but also with grace and faith. Rebirth is a common belief. Various models of God have been noted and we will ask if there is a similarity with the Semitic model in the next chapter.

Both the Buddha and Nanak began as Hindus yet each founded a movement which is distinct in various ways. As we noted, Hinduism is known for its tolerance which springs from its wealth of religious practice and philosophical speculation. It is well expressed by Krishna in the Bhagavad-Gita:

> Whatever form a devotee
> may seek to worship in his faith
> I myself ordain that to be
> in everyone unswerving faith.

It is an attitude which the Semitic religions that we turn to now have not always endorsed.

8 The Semitic Tradition

The three great Semitic religions look back to Abraham as their father. In this chapter we consider the Semitic religions: Judaism, Christianity and Islam. Christianity had its roots in Judaism and Islam recognises the Hebrew prophets and Jesus. God is creator, sovereign, and his revelation must be obeyed.

JUDAISM

The Jews believe in the unity of God (Shema. Deut.6.4–9), and in his action in history. He has a purpose for the world according to the various writers who edited the oral traditions after the Exile (586–538 BC). God has various names: Abraham called him El-Shaddai, the divinity of the mountains, but in the Torah (first five books of the Bible) the names are Yahweh and Elohim. The latter has the plural form though meaning one deity and Yahweh is a personal name. The gods of other nations used Elohim but it is claimed that the Hebrews did not derive it from them. He is depicted as creator, warrior, ruler, guide and shepherd, with a clear distinction between him and Israel. Goddesses are rejected probably because of the Israelite opposition to the fertility cults of Baal current in Syria-Palestine. They were localised but Yahweh is comprehensive and universal. Elohim and Yahweh are in tandem in Deut.6:4: 'Hear, O Israel, Yahweh your Elohim, Yahweh is One'. Yahweh reflects passion and zeal and Elohim means that the Godhead is comprehensively divine, all there is to deity.[1] There is a developing idea of God from the image of a warrior God who is thought to act in an immoral way at times (1 Sam.15.3) to a morally perfect One who is so holy that his name is unutterable. It was replaced by Adonai (Lord) during the Exile. A vision of him results in a feeling of uncleanness and need for cleansing (Isa.6).

Judaism originates from Judah, the sole kingdom after the fall of Samaria in 722 BC. In the eighth century prophets there is the cosmic God, transcendent, majestic, creator, sustainer of the universe (Isa.42.5) and controller of the nations. He creates the stars, and presides over the nations, but he is personal in that he keeps his promises and gives power to the faint hearted. The cosmos consists of heaven, earth, and a region below the earth: the three-decker universe, not to be understood literally. The heaven of heavens cannot contain God (2 Ch.6.18) who elects Israel to be his people, seals the relationship by a covenant and is involved in the great points of their history. Israel knows Yahweh by revelation which brings the message of his righteousness (Isaiah), loving kindness (Hosea), justice (Amos) and new covenant (Jeremiah).

According to rabbinic interpretation (midrash), the Torah has higher and lower meanings which resemble Sankara's levels of truth. In addition to the written Torah God provided Moses with an oral law or explanation which has been passed on from generation to generation. New insights are introduced on that basis.[2] The election of Israel by God did not mean favouritism but responsibility and his punishment must have made them wish that he had chosen someone else! The Jew has a sense of humour and jested that Yahweh did offer the Torah to other nations but they rejected it so Israel was his last hope![3] He protests about his suffering. Traditionally it was viewed as punishment for sin but that did not explain the suffering of the innocent. The question is raised in the book of Job and reappears in modern times.

Yossel Rakover, reflecting on the holocaust, writes about the Jew as a fighter, a martyr and a saint, one that has been chosen by God. He will believe in God even though God has done so much to stop him believing: 'I bow my head before your greatness, but I will not kiss the lash with which You strike me. … I should like You to tell me whether there is any sin in the world deserving of such a punishment as the punishment we have received? … You assert that you will yet repay our enemies? I am convinced of it! … I should like You to tell me however – is there any punishment in the world capable of compensating for the crimes that have been committed against us?' He asks when God will turn his face to them again and what are the limits of his forbearance in the light of such inhumanity: 'Do not put the rope under too much strain, lest, alas, it snaps!' The suffering was too much for many who lost their faith in God.[4]

Some Jews began to understand that God participated in their suffering. Mystics advanced the idea that when the Temple was destroyed the Divine Presence or Shekhinah was separated from the Godhead and went into exile to suffer with them.[5] In the picture of God presented by Deutero-Isaiah, Israel is destined to be a 'light to the Gentiles' and to lead nations to the truth. She failed and had to be punished but it was a purification and a vicarious sacrifice. The servant songs as applied to Israel meant that such suffering would cause nations to experience the salvation of God. The suffering of God was developed further by Christianity in understanding the servant as prefiguring Christ.[6] Jewish thinkers believe that in the light of the holocaust God's omnipotence needs modification. The creation of a free creature limits such power and involves suffering for God. The autonomy given to mankind becomes a reason why he did not intervene at Auschwitz. His omnipotence is not a force which compels us to believe or crushes opposition.[7]

Yahweh is a king, judge, father, shepherd, husband and friend. The images reflect both transcendence and immanence: he dwells in eternity but also in the heart of the contrite person and he treats Israel as a father (Hos.11.1:4). But it is a holy fatherhood and thus different from the human (Isa.1.4:5). Mankind is created in the image of God (Gen.1:26) though it has been marred by sin. God is righteous condemning the guilty, but showing his grace in clearing the innocent and defending the weak against the oppressor. Righteousness is connected with universal salvation as is loving kindness (chesed) (Isa.40:55). The Spirit of God (ruach) is active and broods like a bird over the formless matter before the cosmos emerges out of chaos (Gen.1:2), moves like the wind, and is the breath of God by which mankind lives (Ez.37). It is the agent of prophecy (Ez.37) and of special powers in man (Num.11). As the giver of gifts both technical and spiritual it is a synonym for God. In its brooding, ruling, speaking, quickening, the Spirit is personal.

Man (adam) is linked with the soil (adamah) from which he comes. He is flesh (basar), transitory and frail, but God and the angels are spirit (ruach). The discontinuity is balanced with the continuity for humanity is in the image of God (imago dei) and can commune with Yahweh. The relationship was disturbed by the creature grasping for the knowledge that would make him independent of the creator and elevate him to god-like status (Gen.3:5). The soul (nephesh) is the totality of his being (flesh and spirit), hence the Greek distinction between form and matter or mind and body is not in Hebrew thought. The personality is an animated body, not an incarnated soul; and the spirit of man means his emotions, will and mind, which are not separate from nephesh but the expression of it. There is hope of immortality but in a weak form. The belief probably arose about 165 BC in Sheol (Greek Hades): place of the shades or weak ones (rephaim). A fuller hope is in Daniel (12:2), written in the Maccabean period, and could mean the resurrection of the righteous person. In the Psalms, death cannot interrupt communion with God and there is reference to a rather mundane heaven in the pseudonymous apocalypses (Enoch 21.22.11; Esdras 7.36:). In the first century the doctrine of the resurrection of the body was denied by the Sadducees, accepted by the Pharisees, and modern Jews still debate the question.[8]

In the course of its history Judaism was affected by various influences. The Jews of Alexandria spoke Greek and the philosopher Philo (20 BC) taught a Platonist version of the Faith. God is ineffable, scripture is interpreted allegorically, and anthropomorphic descriptions of Yahweh are kept to a minimum.[9] Maimonides (1135–1204) made mysticism prominent and the deity abstract and impersonal. While his emphasis was more intellectual than the mystical Cabala or Kabbalah (handed down) both stressed the ineffability of God. In Palestine the rabbis did not lose the anthropomorphic portrayal of God for he engages in decision making and suffering, and accepts the prayers of the faithful. Greek philosophy could be used but not as a determining factor in understanding him.[10] The Cabala speculation leads to agnosticism about God. It is reflected today in the way some theologians interpret Exodus 3.14, where Moses enquiring about God receives the mysterious answer: 'I am who I am'. Some assert that it means that God's nature is known only to himself.[11] But a better interpretation is that I will be there as the One who will be there: the leader of his people. God's reply to Moses indicates an experimental knowledge: it is in the experience of Exodus that God is known for the event declares him. His existence is demonstrated by his acts in their history. Exodus 6.2:7 teaches that his name is Yahweh which is related to the verb hayah or hawah resembling the Greek ginomai meaning 'to be' and 'to become or come about'. The 'to be' is not used in the abstract sense of existing in oneself but in the relational sense of being there or being present. Being is not in conflict with becoming but included in working, doing and acting so we can say that the scripture teaches the Being of God but understood in an active sense which opposes the Brahman static concept or the unmoved Aristotelian mover.

We need to be careful in speaking of God as impersonal for this is to neglect the dynamic agent model. Perhaps those passages which speak of his glory and majesty point to his spirituality rather than impersonality. But there are warnings about making God like man: the forbidding of graven images. The vision of Ezekiel depicts God in human form: 'the likeness of a human appearance' (Ez.1.22, 26:28). Fire and brightness do mean glory but the humanity embarrassed the Jews so that the Mishnah (oral law) commands that the account will not be read in the synagogue. God is incomparable: 'To whom will you compare me that I should be like him? says the Holy One' (Isa.40:25). But the scripture uses anthropomorphisms which is the way God accommodates himself to us and are not to be taken literally. They show an affinity with mankind but also a difference. When we read about his concern, graciousness, slowness to anger, faithfulness and disappointment concerning Israel's failure, forgiveness and willingness to change his intentions and judgement, we realise that God's humanity is vastly superior to ours. God is not like sinful man for he does not lie (Num.23:19) or act as a destroyer (Hos.11:9). It is humanity as it should be. The contrast becomes even sharper in the Christian incarnational model of God, for Christ 'turns the other cheek', shows compassion, suffers for others, is servant-like, loves and forgives his enemies. Human images of fatherhood and motherhood point to the divine ideal and in that sense are indispensable.[12]

The Hebrews thought that God had human characteristics for it is inconceivable that we could see and hear and not God (Ps.94:9)! God and man are not incompatible for we are made in his image (selem) and likeness (demuth) (Gen.1:26f), which marks the distinction from the animals (Gen.1.26:27). But he is in continuity with them, being from the dust of the earth. Thus the scripture is sometimes optimistic about human nature (Ps.8), for 'he is a little less than God' and at other times pessimistic (Job 14). The image of God in mankind has been defaced, hence when we speak of God in human terms we are not thinking of man as he is but as God intended him to be. The difference between God and man must not be forgotten because of his sin, temporality and sexuality.[13]

The Hebrew believed that God made himself known through his actions. The Exodus was his act as was the Exile from their land (Isa.41.21:29). Israel believed that a God who was unable to act in history did not exist. As G. von Rad points out, through the 14 centuries of biblical history when five great world powers (Egypt, Babylon, Phoenicia, Persia and Greece) dominated, the prophets of Israel constantly discerned the agency of God. But this was not perceived by their conquerors. Pharaoh thought that Moses was some kind of magician and that the Egyptian gods were more powerful than Yahweh. There is no mention in their records of any disaster at the Red Sea which may mean not that it did not occur but that they were too embarrassed to record it! But the actions of God are of an indirect nature which required insight. They can be experienced privately or publicly: Pentecost (Acts 2). When the event happens in this way it is noted but usually a natural explanation is put forward: the public thought the disciples were drunk. At the baptism of Jesus a noise like thunder was heard but the power of the Spirit coming upon him was not observed nor did Paul's companions hear the voice that spoke to him on the Damascus road. There is something hidden about the act of God so that the event requires the spiritual discernment of the observer.

The empirical philosopher basing his view on the observation of causes rules out such invisible factors but in quantum physics things can happen and causes are difficult to detect. A solution has been attempted by David Bohm. In his interpretation of quantum theory he divorces wave and particle instead of thinking of complementarity. We can observe the particles which are objective but we cannot see the wave yet it guides the motion of the particles and encodes the information about the environment in which they move. Thus he argues that we have the explicate order which is observable and is the area of reductionist science, and the implicate order which is veiled and holistic.

What is the relation? The explicate is enfolded in the implicate order which is its ground. Consciousness is an example of it but it has its explicate manifestations in the perceptions of the ego and memory. In brief there are hidden variables which if we could discover them would help us to understand causes in quantum mechanics. While there are objections to Bohm's view because he is postulating hidden variables in nature, his explanation cannot simply be written off. Neither can the theist who by analogy speaks of a hidden variable which he calls God as one way of explaining the extraordinary.[14]

The believer sees a divine pattern or meaning or significance in events whereas the unbeliever speaks of accidents and coincidences. When one of Gandhi's sons fell ill with typhoid the doctor recommended eggs and chicken broth but Gandhi was a strict vegetarian and decided to try hydropathic treatment instead. He knew that he was playing with his son's life and was assailed with doubts but persisted in his course in the hope that 'God would surely be pleased to see that I was giving the same treatment to my son as I would give myself'. He also reflected that 'the thread of life was in the hands of God. Why not trust it to him and in his name go on with what I thought was the right treatment?' It was night and he left his son in the care of his wife and went for a walk to refresh himself: 'It was about ten o'clock. Very few pedestrians were out. Plunged in deep thought, I scarcely looked at them: "My honour is in Thy keeping oh Lord, in this hour of trial", I repeated to myself. Ramanama was on my lips. After a short time I returned, my heart beating within my breast, and found that the fever had broken.'

His son made a full recovery. Afterwards Gandhi posed the question whether the recovery was due to God, to hydropathy or to nursing and was sure that it was due to God. Gandhi saw God not only acting in this way but also in the Indian independence movement.[15] Since God acts indirectly through human response and cooperation we would say that the son recovered with God as the primary cause and the treatment as the secondary, that is, the natural one. The former complements the other without contradicting. But what if the prayer for recovery had not been answered? Some argue that it shows that God does not or cannot act but others contend that it is the will of God which will ultimly be revealed. Such patience to wait for an inactive God is vividly portrayed in the book of Job where friends and wife complain and call on Job to curse him. But Job's faith scales the heights: 'though he slay me yet will I trust him' (Job 13.15). Without faith there can be no discernment of the act of God.

CHRISTIANITY

At first the Christian faith was part of Judaism but as it took root among the gentiles it spread rapidly through the Roman Empire and though constantly persecuted became the official religion under Constantine in 313 AD. The Western Empire fell in 476 AD but Christianity went on to convert the barbarian invaders. The concept of God in Christianity derives from its Jewish roots and the teaching of Jesus. The Christians accepted the revelation of God through individuals, law-givers, judges, priests and prophets but contended that such revelation culminated in Jesus Christ who was the Son of God (Heb.1.1). Christianity is a monotheism with an emphasis on the personality of God: he is Father.

Christianity is based on the life, ministry, death and resurrection of Christ. The first three gospels present a common picture of him and the fourth engages in theological and mystical reflection. The main facts are that Jesus was a Jewish peasant who came from Nazareth and had an active ministry of no more than three years during which he lived the life of a wandering teacher. He incurred the hostility of religious leaders and was executed by the Romans on a charge of treason. On the third day after his death his followers asserted that they had seen him alive and that he would return to them and set up his kingdom. No other religion attaches so much importance to its historical founder, hence the importance of the records which have been sifted by historical criticism.[16]

Christianity is theistic: a belief in a transcendent being with properties such as omnipotence, incorporeality and personality. God is Father in a very personal way but he is also transcendent for the Lord's prayer speaks of him as being in heaven. Jesus preached the kingdom of God which was both present in his preaching but was to be fully revealed at the end of the world (Lk.17.20f). Since God is Father his love for the sinner is depicted in many parables and Christ dies as a priest who offers a sacrifice for sin. The early Church saw the cross as the proof of God's love for us (2 Cor.5.19). Jesus exercised a healing ministry in which suffering was regarded as an evil that must be defeated (Lk.11.14f). Most Jewish and Christian exegetes would accept that Jesus was a preacher of the kingdom of God, a charismatic figure, and one who performed acts of healing but there is more doubt about the nature of miracles.

As in Judaism the acts of God through Jesus were not discerned. The religious leaders doubted his miracles or asserted that he performed them by the power of the devil. What they worried about was his effect on the masses so they plotted to remove him. Jesus points to their closed minds in the parable of the rich man, Dives, and a poor man (another Lazarus) at his gate (Lk.16.19–31). In the afterlife their positions are reversed and Dives, suffering for his sins of omission concerning Lazarus, begs Abraham that someone be sent back to earth to warn his brothers of the fate in store for the uncharitable. But Abraham says to him 'If they do not hear Moses and the prophets, neither will they be convinced if someone should rise from the dead' (Lk.16.29–31).

What is the relation of this insight to reason? Is insight another name for blind faith as Richard Dawkins would say? The Bible commends reason for it tells us to love God with our minds. His thoughts are higher than ours but this does not mean that they are completely different, otherwise there would be no point in the request: 'Come now, let us reason together, says the Lord ...' (Isa.1:18). The people who came to Jesus were Jews who had the Torah and the traditions and critically listened to what he said. As Kant wrote: 'Even the Holy One of the gospel must first be compared with our ideal of moral perfection before we can recognise him to be such.'[17] There are few signs that Jesus was a dictator or a kind of religious virus, as Dawkins insists, that operates without the consent of the receiver. Jesus recognises that people were capable of sorting out the good person from the bad in his parables (Lk.10.36). In his debates with the religious rulers he defeats them by reason and he tells those who wish to follow him to consider carefully the cost of following him and many refused to pay the price. Some with more spiritual insight perceived who he was. Peter confessed: 'You are the Christ, the Son of the living God' (Mt.16.17). Jesus said that it had been revealed to him by God. Such occasions were rare for Peter, however, since afterwards he fails to understand the suffering of Christ and is rebuked (Mt.16.22). This kind of insight is connected with the grace of God and differs from the self-effort of intuition or imagination. But if seen as a heightening of human powers some reconciliation can be postulated.

There were varied opinions about Christ reflecting debate (Jo.7.12), and doubt about his miracles (Jo.12.37). Faith is not against reason for it considers what is being said and done and is open to the enlightenment of the Spirit. The people ask for signs and wonders, a spectacular display of his power but Jesus has already refused this in the Temptation narrative. He has no intention of forcing them into the kingdom by dazzling them, miracles will only be done for human need. His death on the cross seemed to confirm their doubts about him for it was the worst fate that could happen to anyone. As Cicero said: 'Far be the cross not only from the bodies but also the minds of Roman citizens'. In contrast, the Buddha, Nanak and Muhammad died peacefully with honour and dignity. The Gospels show the disciples fleeing, defeated and scattered, and record that what changed them was a unique event: the resurrection. In the event we have the agent model of God for it is by his act that we are told Christ rose from the dead. How much evidence is there for this? The accounts of it are surprising and and not what might have been expected. The Jews thought of resurrection in terms of glorious heavenly beings (Dan.12.3; Wisd.3.1.7) or the restoration of someone to his normal human state such as the widow's son by Eligah (11 Kings 4.8–37). But the accounts of the risen Jesus do not fall into either of these ready-made formats:

> This faces us straightaway with the following question; if the disciples were inventing the story that Jesus had come back to life, or if they were simply trying to convey an inner certainty that his soul, like that of John Brown, went marching on, is there not a probability that they would have used,

albeit with minor variations, a convention recognizably akin to those already current in their world of thought?[18]

Another novel matter is the witness by women of the resurrection, some of whom had doubtful reputations. They did not count as valid witnesses in Jewish society. The timing of the event was surprising since resurrection was expected at the end of time and it was bodily not the immortality of a soul. The latter would not have needed an act of God for immortality was an inherent quality of the soul. Such an act of God cannot be detected by the historian but its effect assessed: the change in the disciples. Some scholars conclude that such an event is beyond proof or disproof but scripture records the appearances of Jesus by 'many convincing proofs' (Acts 1:3) and Paul produces witnesses to the event (1 Cor.15; cf. Acts 1.21:22).[19]

What of the nature of the resurrection? Was it the dematerialisation of a physical body and its reanimation as a spiritual one? The resurrection appearances show that there was some connection between the old body and the new. The disciples recognised him not by physical appearance but his actions which reflected how he acted in his earthly life (Lk.24.30:31). We know that change in our physical appearance can prevent recognition. When we see someone after a long time who has changed physically we may not at first recognise him or her but then an action, how the person walks or speaks, suddenly induces recognition. In order to establish a connection between the old and the new body, Paul uses a number of analogies. One is the seed and the plant: the seed sown is related to the new plant that sprouts but the new sprout has a different body. Our bodies die just like the seed but there is resurrection with a new spiritual body, hence Paul says: 'flesh and blood' cannot inherit the kingdom (1 Cor.15:50).

The soul or self or personal identity could be replaced by a new one like the software in a computer or replicas/copies could be made. It would be the transformation of the body not the reassembling of the old atoms but the pattern inserted into a spiritual one. What does seem clear about the resurrection of Christ is an identity yet difference which could be important for the relation of God and matter. Let us assume that the resurrection took place and the stories about it are trying to grasp the inconceivability of the event. The stories of the mode of the resurrection can be compared with how Ramanuja sought to describe God and his relation with us: identity-in-difference. If Jesus was God incarnate then his resurrected body is an example of what God might be like in the personal aspect of his Being. The body belonged to the divine dimension or order, so unlike ours, yet it was used in our time and space. On the other hand, it was not confined to one location or space: transcended our dimension. The story of the transfiguration, which some scholars think belongs to post-Eastern sources, can also be seen as a transformation of the earthly body and the effect on the disciples was equally devastating. What we can say is that such experiences show that God is related to matter and to our time and space in such a way that it is not alien to him.

The act of God in the resurrection of Christ fits into the pattern of the acts of God which we have seen in Judaism. God grants liberty to man and the world but his justice operates when freedom is abused by evil. Thus he acts in the Exodus from Egypt to liberate the Hebrews and subsequently throughout their history. When trust is placed in him he responds and evil doers pay: death was the penalty for the Nazis who created the concentration camps and a devastating destruction of Germany. A new future is opened to the Jews in the state of Israel. With regard to Jesus it was not just that his trust in God would end in failure or that he should abandon him. Death could not be the end. His identity also needed to be revealed and it was the resurrection that confirmed that Jesus was the Son of God (Rom.1:4). But it raised the question of how someone of 'the seed of David according to the flesh' could be both divine and human (Rom.1:3). How God could become man became a central issue in the first centuries of the faith and resulted in the doctrine of the two natures and the Trinity. These models of God will be discussed in a later chapter.

A new image of God is seen in Christ (Col.3:10). It is he who has the true image because Adam failed to preserve it. Hence when we speak of God in human terms it should be not in terms of sinful man but as God intended him to be. The Adamic inheritance meant that the creature was unable to see God in creation or Christ as the revelation of God. Man had been made a little lower then the angels but grasped at equality with God. Such disobedience was reversed by Christ, the second Adam. A new creation is necessary (2 Cor.5:7) and gradual changing into what Christ is like by the Spirit of God. But in general any pessimistic view of our nature is balanced both in Judaism and Christianity by an optimism concerning natural goodness (Ps.8; Lk.29:37). We are like God (Gen. 1.26:28) but unlike him (Isa.40.18:25).

ISLAM

The central figure in Islam is Muhammad, regarded as the final prophet. It is difficult to know what the influence of Jews and Christians was on Muhammad but the 'people of the book' did attract him and it was only subsequently when they reacted in a hostile fashion that he returned the compliment! Born in 570 AD he never learned to read or write, hence 'the unlettered prophet', but he heard the stories of the Bible from the Jews and Christians whom he met. He had originally been a shepherd and had the natural inclination of the prophet or mystic to withdraw from the crowd and spend time meditating in the caves of Mount Hira. The monotheism of the other Semitic religions was attractive when compared with the beliefs and customs which he saw around him. Each Arab tribe had its own god, worship was localised, and superstition was rife. But there were no idols in the sense of images or representations. Allah (al-ilah), the name appropriated for the God, was a title which a tribe conferred on its own particular god therefore it was useful for unification of belief. Muhammad was grieved by the false notions about Allah among the Arab tribes, the constant internecine strife, and the clear evidence of immorality.

He received visions and was told to recite what he had heard and these utterances of Allah became the Muslim scripture. Apart from Muhammad's sensitivity to such visions and his undoubted charisma, the question of how such revelations are received has been debated. This applies not only to Islam but other religions. The Muslims hold that it was due to Allah alone with Muhammad as a passive instrument, but does God not reach the human by heightening the powers of perception? Muhammad struggled to understand the words. Does this mean that the human can misinterpret revelation and make mistakes?

After passing through a stage of doubt Muhammad survived conflicts with the Meccans, overcame them, and founded a religion which spread with great rapidity. His greatness is shown in his endurance of ridicule, threat and persecution, leadership, and steadfast belief in his prophethood. But while he could be merciful he was also harsh at times: he ordered the beheading of the Jews because they had supported the Meccans. Religious insight and political power combined in the success of his mission. The concept of jihad or holy war is the duty imposed by the Qur'an on Muslims to fight against those who believe in many gods. But there is the Sufi, which understands jihad as the spiritual battle against sin. Muhammad followed not the suffering tradition of the Hebrew prophets or Jesus but Moses and David who fought wars.

Medina became the base where Muhammad could plan his strategy and issue forth to fight the decisive battles of Badr and Uhud and plan the conquest of Mecca. Muhammad never made any claim to be other than a messenger even though among certain groups of Shi'ah he is thought of as an incarnation. The verse that was revealed to Muhammad, after the battle of Uhud, was quoted by Abu Bakr on the death of the Prophet:

> Muhammad is naught but a Messenger; Messengers have passed away before him. Why, if he should die or is slain, will you turn upon your heels? If any man should turn about on his heels, he will not harm God in any way; and God will recompense the thankful.[20]

The verse was appropriate for those Muslims who at the battle of Uhud had thought that the Prophet had been killed and were considering fleeing.

Some have alleged that Muhammad was an epileptic, a charge levelled at the apostle Paul and others who have 'seen visions and dreamed dreams'. But when we consider his common sense and leadership qualities, his shrewd appraisal of others and his ability to sense what was going on in the world, his persistence in the face of opposition and his unification of his people, it is clear that the allegation is false. Muslims believe that he is an example to be followed as an extraordinary man, a precious gem among stones, and a sign of the presence of Allah. Hence the Muslim reaction to the Rushdie affair.[21] The success and rapid spread of Islam was due to the personality of the Prophet, the need which it met, the high moral tone of the message, its monotheism and the force which it displayed. All three Semitic religions look back to father Abraham but the Muslim sees the promise of God to make a great nation as referring to Hagar and her son Ishmael and from which they trace their descent rather than the Jewish claim that the reference is to them through Sarah and Isaac (Gen.16.11). The Muslim bears witness that there is no god but Allah: and Muhammad is the messenger of Allah (Shahadah). Allah is self-sufficient, he does not beget, neither is he begotten, hence incarnation is impossible since Allah cannot be diminished or humanised or become like a creature. He is transcendent, is Truth (al-Haqq), Light (al-Nur), the Witness (al-Shahid), the Seer (al-Basir), the Hearer (al-Sami) and the Creator (al-Khaliq). He is omnipotent but also compassionate with his mercy taking precedence over wrath. As judge he will raise the dead to life and reward or punish mankind according to their deeds.

The desire to make Allah transcendent has its problems and is seen when 'the beautiful names' of the deity is commented upon by Muslim scholars. Thus at times he is described in a human way. The Hadith or collected sayings of the Prophet stresses the presence of Allah in the believer so that he becomes the ear through which he hears, the eye with which he sees, the hand with which he grasps and the foot whereon he walks. Immanence balances transcendence. The study of the names of Allah shows a close relation with humanity: he is the life-giver, the restorer, the subduer, the fashioner, the listener and seer, the forgiver and the friend. Reflection on these names by the Muslim might lessen his antipathy to incarnation.

Islam has been accused of fatalism for Allah is sovereign, yet the Qur'an insists that when we fall into sin it is our fault. These has been much debate about free-will and predestination in Islam but the difficulty is that the Qur'an is not clear and some assert that belief in fatalism reflects pre-Islamic ideas. What we do know is that passivity of thought and action occurred in

Islamic life during periods of decline such as in the European conquest of their lands and may have encouraged a fatalistic belief. Allah is Lawgiver and Lord rather than Father or Shepherd and his rule extends over all areas for there is no division between sacred and secular. Allah is merciful, generous and forgiving, and he has breathed his Spirit into mankind. He is absolute and eternal and alone deserves worship.

Man is weak and ignorant, and easily falls into sin, but he is not totally depraved for he has capacities for good as well as evil. Allah has sent messengers to all people to guide them into the right way (Surah 35). Before Muhammad these messengers brought partial truth about Allah but he, 'the seal of the prophets', corrects and completes the structure of revelation and provides a final synthesis after which there is nothing more to be said. The commandments of Allah must be obeyed and sincere repentance is expected for wrongdoing.

The Qur'an recognises the virgin birth of Christ (Surah 19), his miracles and, unlike Judaism, his Messiahship. Of no other prophet mentioned by the Qur'an is this the case not even Muhammad. It is recorded that Mary said to Allah: 'Lord how can I bear a child when no man has touched me?' and the reply is, 'Such is the will of Allah, he creates whom he will. When he decrees a thing he need only say: "Be" and it is.' The miracles of Jesus are accepted, not only healing but raising the dead to life. While Allah cannot beget a son, Mary's child is 'a holy son'. But the resurrection of Christ is denied. The Jews thought they had put him to death, but 'they did not kill him, nor did they crucify him, but they thought they did. Those that disagreed about him were in doubt concerning his death, for what they knew about it was sheer conjecture; they were not sure that they had slain him. Allah lifted him up to his presence …'[22] Islam puts foward a test of the true religion: it is one that accepts the messengers of God. The Jews rejected the prophets and Jesus and Christians reject Muhammad but the Qur'an accepts all the prophets hence it is the true religion. But such a test fails because many Jews accepted the prophets and most Christians today would accept Muhammad as a prophet.

However, the high status given to Jesus does not make him a son of Allah. This may be due to thinking literally about physical sonship or reaction to the old gods of Arabia. Today Christian scholars have tried to assure the Muslims that such sonship means the generation of God's love evident in Christ or that it means that Jesus was obedient to God as Muhammad was. But the traditional doctrine means more than this for it resulted in the doctrine of the Trinity. There is also the Muslim denial that Jesus suffered physically on the cross probably because the Arabs think of a prophet being successful in his mission. The Muslims believe that Muhammad was endorsed by Jesus for the Qur'an says that he spoke of an apostle who would follow him (Surah 61). It is also affirmed by them that Jews, Christians and Muslims worship the same God: 'Be courteous when you argue with the followers of the Scriptures, except with those of them who do evil. Say: We believe in that which is revealed to us and which was revealed to you. Our God and your God is one. To him we surrender ourselves.' (Surah 29).

Despite the Muslim creed: 'Allah is one, the Eternal God. He begot none, nor was he begotten. None is equal to him', the Shi'ah who represent about 15 per cent of Muslims developed incarnational ideas in connection with Ali who was the cousin of Muhammad. They believe that he was the first true Caliph instead of Abu Bakr who succeeded the prophet. Shi'ah is the official religion of Iran and its members encountered great suffering. They have the concept of the hidden Imam, mysterious, invisible, immune from death, infallible and impeccable, and allegiance to him is regarded as a sixth Pillar of the Faith. The belief arose when their 12th hereditary Imam died without issue and his place was taken by the hidden Imam. They reject the consensus (ijma) whereby official Islam, the Sunni, identify development in the law and usage. The Sunni think of success as the criterion of the prophet, hence there is no thought of a suffering Allah but the death of Husain, one of the sons of Ali in 680 AD in the massacre of Karbala (Iraq) due to his bid to be Caliph, made the Shi'ah dissent from such a criterion and impelled them to believe in vicarious suffering. Some elevated him and his descendants to a near-divine status and thought of the successive Imams as avatars.

This was denied by the Sunni (83 per cent) but they had also to deal with the speculation of the mystic Sufism (two per cent with the Islamic brotherhoods) who taught that Muhammad was the perfect image of God. The name Suf (wool) derives from the early exponents who wore a woollen or serge dress similar to that of the monks of other religions. It was a movement to find an approach to God other than external practices and legal justice: a search for a personal and intensive experience of religious truth. The Sufi point to Muhammad with his mystical visions and his account of the miraj or ascent to heaven where he saw the prophets dwelling in the various heavens. Mystical meanings could be found in the Qur'an and the five pillars of Islam need not be interpreted literally. The divine unity could be seen as the Real in whom we are and have lost our individuality, and their rejection of the self resembles Buddhism. Some members of Sufism such as Ibn Arabi (1165–1240 AD) of Spain moved to pantheism, regarding the world as a visible manifestation of the divine Reality with mankind as an offshoot from the divine essence. It was due to a gradual development under Sufi influence that an aura of more than human honour was built around the Prophet who became the emanation of heavenly light and the cosmic repository of all truth. In so doing they are exhibiting the incarnational strain which is anathema to the Sunni mind.[23]

When Muhammad died Abu Bakr and Umar offered this prayer for him:

> Our Lord, we bear witness that he has conveyed to us what has been revealed to him, given good counsel to his nation, struggled for Allah's cause until Allah has given triumph to his religion at his hands, and until Allah's words were complete. People believed in him alone without partners. Our Lord, place us with those who follow the word revealed to him, and join us to him so that he recognises us and you make us known to him. For he was compassionate and merciful to those who believed.[24]

CONCLUSION

The three Semitic religions understand God as having an overall purpose leading to the consummation of all things: the eschatological model. The Jews saw God leading them into a new future: the new Exodus, the new Jerusalem, the new covenant, the new creation and consummation. The Muslim looks for a Paradise with rivers of milk and wine, and those who reach it will wear robes of silk, recline upon soft couches, and be attended by boys graced with eternal youth. Such a vision inspired Muslims to go into battle, for death meant immediate entrance to it. In the final assessment all who die recover their bodies and stand before the Judge and to each is passed a book. If it is placed in the right hand it is the passport to Paradise, if in the left it means hell which is described in vivid imagery. But if Muhammad prays for some of them they can be redeemed. Christianity has its 'new Jerusalem' with its ending of strife and war and the healing of the nations. Buddhism speaks of an ultimate nirvana, and in the Mahayana tradition there is the blessed land.

In contrast with the optimism of the religious view the scientific picture of the end of all things is bleak, for the second law of thermodynamics shows that matter becomes increasingly disordered: a descent into chaos ending in a 'heat death'. Bertrand Russell, believing that the universe was the result of an accident, complained that all the labours of the ages, all the devotion, all the inspiration, all the noonday brightness of human genius were destined to extinction in the vast death of the solar system. But religion asserts that while present pain and suffering, injustice and tragedy are real, it is not the last word. The view of God in the traditions is One of limitless love and grace who will turn tragedy into triumph and make life's journey worthwhile. While this vision of God cannot be proved or disproved it corresponds more to our natural feeling of justice and fairness.

Eschatology is often thought of in this way but it is contended that God's future purpose has been fulfilled in the lives of those who have been obedient to him. Jesus, for example, has been described as the eschatos, the one in whom God's goal was realised and the Muslim and the Sikh would equally see God's purpose being fulfilled in Muhammad and Nanak. Hence eschatology is not simply to be equated with a future consummation but with a God of the future continually operating now in the history of mankind.

When we consider the Semitic with the Indian religion we can stress the differences: the lack of belief in God in early Buddhism, the historical emphasis on events in the Semitic and the grace of God as opposed to the philosophical and mystical approach of the Indian. There are also differences regarding the soul, the identification of it with God, and reincarnation. We do not minimise these differences but would point out that there is diversity within traditions and sometimes comparisons are made between extreme forms of a faith. In particular, to say that prophetic religion starts with God and his dealings with man in contrast with the mystical tradition which begins with man's efforts to reach God is too simple a generalisation, for there is mysticism in the Semitic religions: St John of the Cross, Eckhart, Boehme in the Judaeo-Christian tradition and Sufism in Islam. And, can the experience of the external be separated from the internal?

Apart from Buddhism, which does not recognise a creator, there is a belief in both the Semitic and Indian religions that the universe is dependent upon God. He is involved in the world as the compassionate and merciful Allah, or as the Yahweh who acts on behalf of his people, or as Vishnu who is incarnate because of the need of mankind, or as God loving the world to such an extent that he gives his son. There is a combination of the transcendence and immanence of God and there is a theistic element in most. The law is prominent in Islam and Judaism with obedience and submission as key factors but grace and help from God is proclaimed in Sikhism, Christianity, Mahayana Buddhism and in the Hinduism proclaimed by Ramanuja.

All three Semitic traditions and Sikhism treat the world as real: it can delude us but it is not an illusion. In contrast to the Indian tradition, the historical process is linear not cyclic and revelation is in event with God active not static. Union with God is conceived not in terms of identity but the closest of relationships. The personal nature of God emerges in most. The six religions have much in common ethically: what God requires of mankind. Some put greater stress on the impersonal aspect of God but we would contend that the personal is the highest aspect of humanity, and should be given primacy.

The Indian religions, apart from Buddhism, see some sort of continuity between God and mankind. Judaism and Christianity speak of the image of God in man and the continuing of the relationship through covenants. The avatars of Hinduism and the doctrine of the incarnation in Christianity show that divinity and humanity are not alien to one another. Islam speaks of the closeness of Allah to the Muslims and his compassion towards them. But we also noted marks of discontinuity in the Indian traditions between man and God and the same applies to the Semitic. It is shown in disobedience or ignorance which are related for we cannot obey if we do not know. The prophets of Israel lament the lack of knowledge of Yahweh and Christ accused his enemies of not understanding what God was really like. There is also the doctrine of the Fall of mankind in Christianity which means a nature opposed to doing the will of God. Islam, while agreeing that we are imperfect, sees sin as an act of disobedience and not due to an Adamic inheritance. There is also diversity within each religion so that the depravity of mankind which surfaced in Christianity agrees with Madhva's view in Hinduism. But whatever the differences they all in various ways would deny the scientific reductionist claim that we are simply machines.

Various models of God have emerged: the dual, incarnational, self-limiting, transforming or emptying of God, being and becoming or being-in-action, agent, God of the future, and the various founders of religion are seen as examples or models of what God requires. We will continue to probe these models in the next chapter and ask if there is any similarity with the way the scientist uses models to picture the world.

9 Models of God

There can be no comparison between God and earthly things
but the weakness of our understanding forces us to seek
certain images from a lower level to serve as pointers to things
of a higher level. Hence every comparison is to be regarded
as helpful to men rather than suited to God since it suggests
rather than exhausts the meaning we seek.

Hilary of Poitiers

We want to consider other models of God in this chapter but
first we look briefly at the use of models of the world in
science. Science has made its great advances by observation,
experiment, and the testing of theories. These go beyond observation and no theory is in agreement with all the facts. It is
naive realism to think that we have in the laws and theories an
exact replica of nature. Science uses models, analogies and
symbols, as in religion, and these are unable to point to all the
features of an object. But they represent realities in the world
and are not just useful fictions. They are neither simply based
on sense observation nor identical with mental states but approximate to the truth of reality. Evidence can be produced,
for example, to show that atoms and electrons are real. But as
in religion, the models are limited and partial and do not describe reality as it is in itself. We know how the world behaves
not what it actually is. Both in theology and science models
are revisable and subject to change.

Concepts are expressed in language or mathematical formulae which point to structures in objects which are hidden
from empirical investigation. Science uses analogy which
relies on a similarity but difference between entities. Thus
the wave theory of light developed on the basis of an
analogy with the wave properties of sound: some of the characteristics were similar but others different.[1] In the same way
when we speak of God on the basis of what we are and how
we act we recognise similarity but difference as we saw in discussing the anthropomorphisms of religion. A model could
be called a systematic analogy because we are moving from

154

some law which is understood to another which we hope to know.

Mathematical models are symbolic representations of physical systems used in order to predict behaviour but theoretical models spring from creative imagination. The concepts of science are symbols and very indirectly refer to the atomic world because there is no direct observation. Of course every discipline is interested in some particular aspect of an object and only describes it from that viewpoint. A musician listening to a symphony marvels at the texture of harmony, form and themes, but a scientist hears it as a set of molecular vibrations. Analogies are used from familiar situations to aid understanding before models are expressed in mathematical formulae. They are subjected to rigorous tests and some are discarded.

Since we cannot give a scientific description of the world which explains its inner nature we can infer that the same applies to the creator. We understand what he is like by his interaction with us. Religion and science are dealing with invisible entities. The believer speaks of God, faith, grace, atonement and an unseen world; while the scientist refers to atoms, electrons, nuclei, viruses, hormones and genes, which are difficult to detect and are usually known by their effects. The meaning of mass, force and entropy is not exhausted by observable properties: pure empiricism is not a sufficient basis for science. Scientific theories, as Karl Popper pointed out, are conjectures based on trial and error not a digest of sense observation. Einstein stressed that whether you can observe a thing or not depends on the theory you use. It is the theory which decides what can be observed.

The existence of God is known by his effect on our lives so the test of faith is behaviour. Images of God are similar to scientific models, as symbolic representations of what is not observable, but they differ from the scientific in calling for personal commitment and values, and have a direct relationship to worship and behaviour. There is an ultimacy which is not reflected in science. Models of God, while they reflect more than human ideals, must at least measure up to them. Various societies embrace different values but at least five are fundamental: freedom, justice, knowledge, wisdom and happiness. A religion which opposes these is suspect. The history of science shows the replacement of one model of the world by another and ideas of God develop in religion: the high moral deity of the prophets in Israel and the movement from the impersonal Brahman to Vishnu in Hinduism. Tests for models of God would fulfil the scientific criteria of rationality, internal coherence, fitting with the data, comprehensiveness and fruitfulness. An image of God as king, for example, has given the impression of coercion and needs modification for it does not cohere with lover, shepherd, husband and so on.

Realists hope to arrive at a limited view of truth. We have no omniscient criterion of truth but, as Popper says, that does not render the notion of truth non-significant, any more than the absence of a criterion of health renders it non-significant. A sick man may seek health even though he has not got a criterion for it. Both science and religion require the creative use of imagination, intuition or insight, as seen in the work of Einstein, Clerk Maxwell, Polanyi and Darwin. In the case of Darwin it is recorded that the thought of natural selection flashed into his mind while sitting in his carriage and there is Kekule's dream of a snake grasping its tail in its mouth which suggested to him the ring form of benzene.

Kuhn stresses the faculty of intuition which he thinks of as a 'lightning flash' enabling an obscure puzzle to be seen in a new way. But it does not occur in a vacuum for it is related to the science of the past. It is in the context of the old paradigm but not linked to items of it as an interpretation would be, there is a transformation. In religion the model of God is seen in a new way but building on the past context. Kuhn uses terminology which is familiar in religious discourse to describe these experiences, that is, 'conversion'.[2] But resistance to new paradigms persists because of a lifelong dedication to an old one and the fact that arguments for the new one are not decisive: Einsteins's resistance to the work of Heisenberg. Thus the Jewish religious leaders in the first century clung to their picture of God, resisting the insights of Jesus, and the Arab tribes initially opposed Muhammad's view of Allah.

Einstein spoke of the 'search for those highly universal laws ... from which a picture of the world can be obtained by pure deduction. There is no logical path ... leading to these laws. They can only be reached by intuition, based upon something like an intellectual love of the objects of experience.' As the great mathematician Gauss remarked ironically: 'I have got my result, but I do not know yet how to get it!'[3] The religions speak of an insight into reality which transcends the material and can only be expressed in symbolic language. But while there are parallels between the language of science and religion there are also differences in accordance with the subject matter. Science does not help me to assess whether a poet or an artist or a musician is a good one or not. A different kind of evidence is required, but there is evidence. Richard Dawkins has a problem here for he appears to admit only scientific evidence.

Science and religion do have different viewpoints and answer different questions which can be illustrated by referring to creation. Both science and the Semitic religions agree that the universe is contingent in the sense that it had a beginning and will have an end. But the doctrine of creation means that all depends on God, it is not about temporal beginnings. Hence it would be unwise to infer directly from the Big Bang to creation which would be making a categorical mistake. It would be to explain it in terms of a cause that would not be scientific: ' ... if our universe had a beginning in time through the unique act of a creator, from our point of view it would look something like what the Big Bang cosmologists are talking about. What one cannot say is that the doctrine of creation "supports" the Big Bang model, or that the Big Bang "supports" the Christian doctrine of creation.'[4]

The doctrine of creation does not depend on either the Big Bang or Steady State theories for it is postulated as an answer to why there is something rather than nothing. God is not some kind of physical cause to be placed on a par with the Big Bang but it is due to his will and love for us.

We have considered a number of models of God in the last chapter and noted that in various ways they seek to preserve the transcendence and immanence of God. What is required is a model which does justice both to the majesty of God and his personal nature. Extremes produce deism, a God who is remote and uninterested, or One so human that he looks like a projection of our desires and wishes. Attempts have been made to bring together the impersonal and personal in God by a dipolar view and we look at this first before proceeding to consider the most personal way of knowing God: the Christian doctrine of incarnation and Trinity.

THE DUAL MODEL

The dual model appears in process theology which sees God as the leader of a community. Emphasis is placed on viewing the world as an entire web of dynamic relationships. An organism has different levels as we mentioned – atom, molecule, cell, organ, organism – with influence between the levels and all contributing to the whole. When this is extended to society we see a unity and interaction with each individual contributing creativity and spontaneity. Internal relations are as important as external. Thus the universe is like an enormous reverberating chamber in which any whisper echoes and re-echoes throughout the whole. A. N. Whitehead examines how a new event arises from a preceding cause but each entity handles it differently and makes a contribution, that is, room must be left for the novel and creative. There is a creative selection of possibilities in the context of goals and aims and this is final causation. It is different from physical causation being more like influence: teacher on pupil or the lover on the loved. God moves the human will from within not overruling but stimulating its powers so that an act is free. His influence is the final causality: the attraction of the good. The cosmos is a unity but the mechanist separates it into parts, omits the impact of the whole, and misses the pattern. We are not some kind of substance with qualities for no philosophy has been able to define this concept satisfactorily. In Whitehead's philosophy being becomes becoming and substance process, and we are persons in relationships.

God is the great persuader, companion, sufferer and leader. He is the ground of order and novelty and contains within himself the order of possibilities and potential forms of relationship. He brings pattern and order out of the process and in this he is timeless being the impersonal structure of the world, passive and unchanging. But he is evolving in relation to us, temporal, fulfilling himself, affected by what happens, guiding and luring us to himself. He is the principal of actuality but influenced by the world and in this respect he is temporal and suffers with it. Causality as a physical force is replaced by personal relations on the model of human relationships. Thus there is the dual aspect of God, one personal the other impersonal.

Whitehead rejects an initial act of creation out of nothing. He stresses continuing creation with God as the initiator of all events, he remains eternal in his character and purpose but evolves in interaction with the world. Both Whitehead and Charles Hartshorne try to avoid the criticism that they are presenting a limited God struggling in the process. The latter says that the deity possesses both relative and absolute characteristics. Being is the abstract fixed aspect of becoming but the stress is laid on becoming. God is prior in status being the primordial ground of everything but he is our fellow sufferer. Hartshorne embraces panentheism but does not think of God as a Person which disagrees with John Cobb who calls God the pre-eminent person in a community of interacting beings.[5]

We agree that God is the great persuader, companion, sufferer, leader and pre-eminent person and unchangeable in his character but there is little mention of God as redeemer or liberator. The priority of status is preserved but looks like first among equals. How knowable is the primordial and impersonal aspect and does revelation really flow from it? There is a vagueness here. Process theology is in reaction to the monarchical model of God which goes too far in its modification of God's sovereignty. It requires a stronger doctrine of divine initiative and moral judgement. Critics of process thought insist that the picture of God is too weak portraying him as a kind of 'cosmic sponge'. While that may be going too far the picture of a God emerging in the process appears too limited and the stress on immanence and the social leads to seeing him as a kind of community leader.

On the basis of the religious traditions a dual aspect doctrine of God can be constructed. But if based on Sankara it is not satisfactory nor does it fulfil the criteria of simplicity since many levels can be seen in Brahman. It has at the highest level a God who is unknowable which Ramanuja denied. Ramanuja accepted a duality but neither he nor Sankara were able to resolve it. It is one thing to say that God cannot be fully known because of his majesty and glory, it is another that his personal revelation does not correspond to what he is like. The visions of God in the religious traditions such as that of Ezekiel or Isaiah or in the Gita include both the majesty and personal nature of God. There are hidden aspects but these do not detract from what is being revealed.

INCARNATE MODEL

We have noted the incarnate model in Hinduism and seen incarnational strains in other religions. Even when a religion starts with little reference to the gods, as in Buddhism, there is a development of personal devotion so that veneration is given to the Buddha and the Bodhisattva in the Mahayana form. There are similarities between the Buddha and Christ yet differences. Both taught that we were self-centred and needed enlightenment, neither tried to explain the world that they saw as transient and temporary, and both lived lives detached from those things which tend to dominate ours. But Jesus did not advocate withdrawal from the world and he was conscious of his need for God. The image of Christ which has predominated is of a suffering saviour but the Buddha after his enlightenment lived serenely, had many followers, was successful, and died peacefully at the age of 80 years. Christ died cursed, despised, a false prophet and blasphemer. The image of the Buddha reflects serenity compared with the crucified Christ, yet suffering is the major factor in Buddhism. Mahayana Buddhism introduced the element of compassion and proceeded to develop the incarnational element in the doctrine of the three bodies of the Buddha.

The Shi'ah also developed the incarnational element as did Sufism and certain elements in Sikhism. Divinity can be used in different senses. In Christianity the divine becomes man but in Hinduism it is not clear that it occurs for avatar means 'coming down': the divine manifesting itself rather than becoming human. Both Muhammad and Nanak are charismatic leaders and have an intense relationship with the divine. With the Buddha there is no relationship with God but he is the spiritual principle of Buddhahood, a channel for Dharma: the Teaching. We might think of the Buddha as the 'window' into Dharma, so that the human body, the physical individual, Gautama, virtually ceases to have any importance. The doctrine in Christianity is not easy to explain and there have been heresies and deviations from the orthodox belief. Socinians and Unitarians fall into this category and one is reminded of the encounter between a unitarian minister and a woman who accused him of denying the divinity of Christ. He replied: 'Deny the divinity of Christ! I would not deny the divinity of any man!'

Paul explains how God became man in Philippians Chapter 2. He says that Christ was in the form of God (morphe) and a thing cannot be in the form of another if it does not possess the essential qualities of that other. The difference between likeness (schema) and form (morphe) is that the former refers to outward appearance whereas the latter is essential being. The schema alters, so our appearance at 17 is different from that of 70! But the morphe which is the unchanging element does not change. According to the passage, Christ had equality with God as a right but did not refuse to let it go. Being (huparchein) means innate possession and form is morphe: divine by nature. He emptied himself and became (gignesthai, not a permanent state) in the likeness of man (schema). What does 'emptying' mean? If the classical doctrine is correct it cannot be his essential divine nature but the surrender of his divine state and glory. The Trinity means what affects one Person affects the others so it could mean change in the Godhead, but what does not change is the character of God: his steadfastness and faithfulness.

There must have been some kind of kenosis for if the Son was omniscient and omnipotent then it is a theophany: he only appeared to be like a man, but such docetism was rejected by the Early Church. On the other hand, if he completely strips himself of these attributes then he is downgraded to the level of a man. In neither case do we have the incarnation of God. Any theory of the incarnation must mean the self-limitation of God, hence not only the divine state was surrendered but the divine consciousness limited, as he entered into a different mode of existence. What self-limitation means is not a loss of power by God but its modification in a creaturely mode of existence. It seems likely that Jesus had a growing consciousness of his relation with God and the powers which he possessed. Thus at the age of 12 he spoke of 'my Father's house' and the record indicates that he grew in wisdom and favour with God and man (Lk.2.41:52). It is difficult to make sense of the Temptation narrative if Jesus did not possess exceptional powers. What would be the point of making stones into bread or leaping from a Temple pinnacle if he did not have the power? 'These are temptations which could only have come to a man whose powers were unique and who had to decide how to use them'.[6] Jesus refused such a sensational and magical display for it would have been contrary to the character of God who does not coerce people into believing. But the power requires the cooperation of the human since Jesus was unable to perform a miracle at Nazareth because of lack of faith. It confirms the view that the power of God is exercised through human agency (Mk. 6.5:6).

The incarnate model reveals the love of God shown by humiliation and exaltation. The exaltation occurs through resurrection and ascension and humiliation by his servitude but suppose we reverse this usual interpretation. Divinity, to our surprise, is shown by servitude, for as Jesus pointed out, he who wants to be the greatest must be the servant of all. Exaltation would then refer to the perfecting of his humanity by obedience instead of its distortion by disobedience: he reverses what Adam did (Rom. 5). Humiliation and exaltation are not to be regarded as successive stages in the history of Christ but as two sides of the work of reconciliation. Thus the divinity of Christ is not to be defined by an abstract and a priori conception of divine nature but in terms of the dynamic concept of humiliation and his humanity in terms of exaltation.[7] Revelation changes our view of power and shows that service should be the key element not domination. Our ideas of power are transformed. Evidence for the self-limitation of God is shown by his surprise at the reaction of Israel (Jer. 3.19.5.7), his lack of knowledge for it does not enter his mind that Israel would worship Baal (Jer. 32.35), they do what he does not want (Isa. 54.15), and his change of mind (Jonah 3.10). This human model shows how like man God is, yet unlike him in his patience and love.

God is love and the essence of love is to give. When we give ourselves completely to another person or strive to realise some goal we feel 'drained', 'exhausted' or 'emptied'. But we do not cease to be what we are by the effort. Since God remains what he is and any self-limitation is voluntary he retains his omnipotence and can use it when circumstances demand. Thus the power of God is shown in the resurrection of Christ and in the control of nature (Job 38.41) and will be exercised in the consummation of all things. In specific areas God has modified his attributes in order to relate to us, which appears a better solution than speculations about levels in the Godhead.

It is contended that God cannot be fully or in every respect represented in human form. What does 'fully' mean? If it means that his divine glory and majesty were veiled, we agree, but would contend that he was revealed in accord with the limitation of the form. What is being accepted is the Sankara idea of levels where philosophy attains priority over revelation so that the lower level of revelation, as witnessed to by the scriptures, is cast aside as the mystic attains to levels of the hidden Reality who dwells in dazzling light. But Aquinas said: 'Wherever God exists he exists wholly.' Otherwise we go back to the subordinate Logos of the early Christian apologists which Arius argued was inferior to God. There is a subordination of the Son to the Father in the New Testament but it cannot mean inferiority. We do not say that a son is inferior to his father even though he shows his subordination by obedience. An analogy is that of the piano soloist playing a concerto under the direction of the conductor. He and the other players are subordinate to that direction but we cannot say that they are inferior: their roles are different.

The Christian view of incarnation has been challenged because of its claim to finality in a relative and culturally bound world. We need to think of the Real or the Absolute at the centre of a circle and all religions on the circumference pointing to its nature. None or all of them can penetrate the nature of the Real for they are human expressions of the unknowable. We may know the manifestations of the Real under the various names of God, Brahman, Nirvana, Sunyata, Allah, but not what the Real is in itself.[8] The proposal uses the Kantian distinction between the thing-in-itself which belongs to the noumenal world and the appearances (phenomena) of this world. This distinction has been criticised. All religions make unique claims. The Buddha said: 'Of all paths the Eightfold Path is best; of all truths the Four Noble Truths is best … this is the Only Way; there is none other for (achieving) purity of insight'. He was critical of those who did not understand enlightenment and his experience made him feel superior to the gods and other men.[9] The buddhadharma (teaching) is the supreme expression of the truth.

Muhammad is the final prophet of the Muslims and the Gita teaches that both the Buddhist nirvana and the Brahman of the Upanishads subsist in Krishna who is personal. In the

Upanishads there are three ways of expressing the Supreme Being: Brahman, Self and Person (purusa) but the Gita insists that Vishnu/Krishna is not only all three but is also beyond all three. The Buddhist conception of nirvana which enters into Hinduism for the first time in the Gita must be subjected to the new God. It is a claim which goes beyond the teaching of Christianity regarding Christ as Son for he is subordinate to the Father. Krishna, according to the Gita, is unique in that he has come with a totally new message of devotion and love.[10] In the light of this, the proposal is unlikely to make much headway.

The kenosis stresses the 'emptying' of God which connects with the Buddhist view that only when we empty ourselves of desires can we experience enlightenment. Jesus emptied himself of self or any desire to be on equality with God and demonstrated the drawing power of suffering. Jesus refused to save himself in order to save others and in so doing showed that he was the man for others and the man for God (Lk.23.29). We return to a further discussion of the divine attributes later.

THE TRINITY

The world with its web of relations could reflect a creator who has internal relations: Being-in-communion or relation, but other religions object to the concept of a Trinity. One of the reasons may be that they are thinking of a mathematical oneness and consequently accuse the Christians of worshipping three gods. In Judaism there are two words for 'one': ehadh and yahidh. In the Shema (Dt.6.5): 'The Lord our God is Yahweh, one', there is the use of ehadh whereas yahidh is employed when referring to the son of Abraham (Gen.22.2; Prov.4.3). Ehadh is the word used to describe the relationship between husband and wife: one flesh (Gen.2.24). The two are united as one. Such a view of unity or oneness does not rule out a community in the Godhead.[11]

Islam gives a high status to Jesus, accepts the virgin birth, and regards Jesus as the Messiah. The Jews deny the sonship of Jesus but they refer to Israel being the son of God: (Ex.4.22:23; Hos.11.1f). But Israel was disobedient whereas Jesus obeyed and became the author of salvation (Heb.5.5b, 7:9). Jewish writings on the Messiah equate the words 'first-born' and 'beloved' and it is quite possible that if Jesus did hear the words at his baptism, 'Thou art my beloved Son', he would recall Exod.4.22. His spiritual redemption of mankind parallels the Exodus from the material bondage of Egypt (Lk.9.31) and the New Testament teaches that God's purpose culminates in Christ for having spoken through the prophets he now in a more intimate way speaks, 'by Son' (Heb.1.1:2). Hinduism speaks of Brahman as being (sat), consciousness (cit) and bliss (ananda) and it has also the triad of gods: Brahma, Vishnu and Shiva. Some writers have compared this with the doctrine of the Trinity but Brahma has almost disappeared and Shiva and Vishnu are separate gods. In Buddhism an affinity has been seen with the three Buddhas of Mahayana Buddhism: the historical, the transcendent with a body of bliss, and the Buddha present in all things: the Absolute or the dharma principle.

It has been argued that the Trinity is a later development. This does not take into account the experience of the earlier followers of Christ of Father, Son and Spirit, and the need to reconcile this with the unity of God. The God of Israel was the God and Father of Jesus Christ and it was in that relationship with God that they saw his power and resurrection. The Spirit who brooded over the creation is in intimate connection with the life, ministry, death and resurrection of Christ and the empowerment of the Church. He is the Spirit of Christ bringing conviction of sin and enabling men and women to confess Christ as Lord. In the scriptures, Christ or Spirit are not transitory or inferior entities as in the manifestations of Brahman. If sin was to be dealt with it required God to do it for: 'the unassumed is the unhealed', hence the temporal manifestations must belong in some way to the eternal being of God.

We are social and may expect it to be true of God but does the Trinity oppose the ordinary experience of oneness? There are examples of unity other than the mathematical: the lover, the loved one and the love bond. If God is love he needs an object to love. In mathematics we can think of an equilateral triangle with three distinct angles: 'one God in Trinity and Trinity in unity'. Each angle is phenomenally distinct from the other two: 'so there is one Father, not three Fathers ...'. Yet the angles are similarly structured and equal ('... of one substance ... co-equal ...'). But the problem here is the use of

substance, as we will see. It would be easy to say that God has
played three roles but the distinctions must be preserved since
certain actions stress that God is Father and others that he is
Son. If God were just Son then who would Jesus have prayed
to? There are unities – man and wife 'one flesh', and father,
mother, child – which transcend mathematical unity. The
Church used the Platonic idea of substance or ousia: the
Being of God underlay the particular persons. Plato believed
that the real was the universal but the Cappadocians rescued
the particulars and gave them priority.[12] They argued that God
is constituted by what the persons are to and from each other
in eternal penetration (perichoresis). God is a communion of
persons and each person has his own distinctive way of being.
An abstract metaphysic of being was replaced by the concrete
particularities of revelation.

It is the concentration on something lying behind revelation
that makes it inferior: the stress on the ousia or substance of
God in contrast with his operations (energeia) in the world.
Some scholars want to see him as unchangeable in the first
sense but changeable in the second but as in Sankara the
latter becomes subservient to the former. As a consequence of
the work of the Cappadocians, it is possible to say that the
Being of God consists of the particulars in relation to one
another. God is like this in his inner being and we know it
because of revelation. There is no duality between what God is
towards us and what he is in himself: 'Hence what might be
called the substantiality of God resides not in his abstract
being but in the concrete particulars that we call the divine
persons and in the relations by which they mutually constitute
one another.'[13]

Relations are important in making us into persons but we
are also aware of our distinctiveness. The roles that we play as
father, brother, sister, son, daughter, mother, employee and
employer identify and distinguish us. We exist in these relationships but is it possible to say that there is an ontology of re-
lationality: things are constituted by their relation to other
things? We can argue that while we exist in a network of relations we are entities which hold the relations together. We
have an awareness of an 'I' which transcends events and is not
simply insubstantial in the stream of occurrences. A piece of
music is made up of tones and each is important for the struc-

ture but it is the effect of the whole that conveys the meaning. And we have noted that the religious traditions generally value the person because he or she has been created by God and redeemed by him. Again, we are what we are not only by relations but by our genes and inheritance. Hence in any understanding of the person a balance must be struck between our relations with others and our genetic inheritance.

But when we ask what God is, the scriptures tell us that he is love. They do not say that he is some kind of substance or ousia. We may work out what human nature is on the basis of what we have said above but it is impossible with God since he does not owe his existence to something else: self-existent. On the relational view the Being of God is Father, Son and Holy Spirit and the statement that God is love is not to say that he has an attitude or property but is an ontological statement about the mode of his personal Being in relation. Hence love interpreted in relational terms is understood ontologically.[14]

The Being of God is interpreted as a communion of persons not something static and abstract and it has implications for the divine attributes. Traditionally the latter were discussed separately from reflection on the persons of the Trinity and the divine nature was defined as omnipotent, omniscient, omnipresent and eternal. It raised the problem of how God could become incarnate. But what we see in the Gospels is a relationship between Jesus and God through the Spirit and it is in that way we understand the divinity of Christ. The question becomes, not how the possession of a divine nature is compatible with the possession of a human nature, but how this relationship of Sonship can be exercised and enacted in the reality of a human life, that is, the obedience of the Son to the Father. Christ's humanity is actualising the intention of God which is to restore the imago Dei: the new humanity of the second Adam. The Fall is the dislocation of relationships both with God and other human beings, a seeking of independence from God and reliance on realising personhood without him. Hence the question of the identity of Christ is not answered by abstract metaphysical statements but by telling the story of Jesus to whom God relates as Father through the Spirit.

If we assume that God became man in Jesus Christ then time is not alien to the eternal, change is not incompatible with the immutability of God nor humanity foreign to divinity. The Son can relate to the Father in the temporal as well as in the eternal for the finite is capable of receiving the infinite (infinitum capax finiti). The distinction between creator and creature, however, is maintained for it is an eternal relationship as signified by the Trinity. Jesus Christ exists in a dual relationship: to God the Father and the Spirit which constitute his divinity and to the human which is renewed because of who he is.

The method is to start with what religions are saying in their own context about God and in Christianity it means what God has done in Christ. Paul started from the traditions about Jesus which he had received and from that point reflected on the kenosis or emptying of the attributes of God. But he went on to mention a plerosis in the obedience of the Son culminating in the mighty power of God: the resurrection, the outpouring of the Spirit at Pentecost, and consequently postulated the cosmic Christ. It makes revelation primary not speculation about the being of God. In revelation there is a unity of purpose, commitment and will of the three persons yet each is distinct, just as in a human family of father, mother and son, where there is ideal unity and yet distinctiveness.

The Trinity has the advantage not only of establishing God as a Being in communion but showing the relation of God to the world in a personal way. Personhood cannot be explained completely by the physical and social sciences, though they can help us to understand our cognitive processes, show the relation of brain and mind, and how we develop as persons in relation to others. We have seen a movement towards the personal nature of God in Hinduism, the personal characteristics that appear in Allah, and in Mahayana Buddhism. But the distinctiveness of the personhood of God is seen by the Trinity for it is a holy one. Perhaps in the light of this we should say that God is supra-personal.[15]

THE ATTRIBUTES OF GOD

Omnipotence

Omnipotence is the power which brought the universe into being (Jer.32.37; Qur'an 67) and reflects the sovereignty of

Allah or God. Philosophers have problems in defining omnipotence and suggest two forms, strong and weak, with the latter indicating control. The weak form which suggests the limitations we have mentioned seems best. Like a father, God influences us by teaching, persuasion and patient concern. He points to ideals of goodness and compassion and suffers with his people. The story of Hosea the faithful husband suffering with a faithless wife is an analogy of the suffering of God in dealing with Israel. Such suffering has drawing power (Joh.12.32). Only if we think of suffering as an imperfection, as the Greeks did, will we avoid predicating it of him. But if I say that God cannot suffer it is limiting his power and denying that he became incarnate in Christ.

The scientific picture of the world as involving necessity and chance points to an openness and autonomy which is the gift of God and we choose freely in a framework of his control. When a balance is achieved between chance and necessity there is room for both God and man to manoeuvre which does justice to the freedom of both. The self-limitation that God has imposed upon himself implies that God has chosen areas where he will not exercise his power such as allowing the world and mankind to develop the potentialities which he has built in.

Impassibility

The Greek fathers of the Church opposed a suffering God hence when they spoke of the incarnation they insisted that Christ suffered in his human nature not in the divine. The same view is seen in Islam where, despite the naming of Allah as compassionate, the general picture is of a law-giver rather than a sufferer. Both Aquinas and Maimonides taught that God cannot be affected by anything especially emotions which were a form of disorder and could lead to sin. But if God is love he is affected by the sufferings of the loved one. Thus he grieves over the disobedience of Israel and Christ weeps over Jerusalem. Modern theologians have not hesitated in following Bonhoeffer who spoke about the pain of God. Moltmann sees the death of Christ as a Trinitarian event with the Father suffering the death of his Fatherhood in the death of the Son.

Suffering is part of life as we see from the structure of the world where evolution involves pain and death in order that self-conscious creatures such as ourselves may emerge. We cannot think of God as excluded from it.

Omniscience

Knowledge and how we obtain it is an ever recurring question. Plato thought it was a process of recollection but the empiricist insists that it stems from experience. On that basis it would be necessary for God to be incarnate to fully share the problems of human life. Greek philosophy did not accept it because it would have involved a body, regarded as evil. Of course I do not need to experience everything to have a sympathetic rapport with the sufferer but it is doubtful that even the most sympathetic can enter into the depths of despair of some people. The Greeks also believed that lack of knowledge indicates imperfection, hence God must be all-knowing. At the moment, like the philosophical debate about human freedom and determinism, the relation between divine omniscience and our freedom continues to elude us, though in a recent survey of philosophical arguments Alvin Plantinga concludes that divine foreknowledge is not incompatible with human freedom.[16] But questions remain: can an all-powerful and omniscient Being know frustration? There seems to be evidence of it in scripture: 'All day long I have held out my hands to a disobedient and obstinate people' (Rom.10.21), and it is in the context of a strong doctrine of predestination. We cannot simply dismiss the question by understanding it metaphorically for a metaphor can make the point more strikingly than a literal statement. What we can say on the basis of the religious traditions is that God, despite the way we try to frustrate his plans, has overall control, and will overcome evil in the end.

Our world has certain systems whose future states cannot be known: the 'Heisenberg' range and perhaps non-linear systems at the macroscopic level. But this does not prevent God knowing the probabilities of the sequence of events in such systems and influencing the general direction of natural events. God would know sufficient about us to influence the direction we take. If the future is not open in some way why pray? Is it not more than psychologically adjusting ourselves to difficult situations? Are we not told that if we have faith our prayers will be answered, perhaps not in the way we expect but in some way? The religious traditions show that God can change his mind when prayer is made.

Immutable

The traditional view was that God was an unchanging ground of change, a Being necessary in order to guarantee stability in a changing world but the biblical reference, 'I am the Lord. I change not' (Mal.3.6), had no metaphysical notion of immutability. It is faithfulness (hesed) to his purposes and is known in his dealings with Israel which show how he changes. Plato's view was that in our world there are things that become or change but they do not possess real being. Conversely in the world of the forms, they can be, but never become. In this context, to be, has an inertness about it: Aristotle's Unmoved Mover. But immutability can be understood as unchanging in his nature, character, laws, justice, love and mercy, which governs the way he changes in response to us.

Change is connected with time and if we think of God as timeless it is difficult to see how he could change or act in time, hence it is better to think of him as everlasting. He creates time with the universe and is closely related to it. The theory of relativity makes time relative to each observer and questions its flow. How can God relate to each observer's time? Does he relate to each of them? We need to think of time in two ways: physical and psychological. There is the physical structure or causal sequence created by God which is independent of us and there is our experience of it. We are aware of the present, remember the past, and realise that there is a future: aware of the succession of our conscious states.

If we argue that God is personal in a higher sense than we then he too must be aware of succession for how can we conceive of a consciousness that does not have succession? We might think of God creating each instant of time which would make the future open for him since he has not yet created it. But God being omnipresent is not located at any particular point of time even though he creates the instant but is spread out all over it, hence he experiences everywhere and everywhen.[17] The totality of experience is in God. All of which would indicate that God is reacting to what we do and therefore will change according to our response. Otherwise we make God a prisoner of his immutability by defining it in a way that prevents him reacting to what we do. Traditionally it was thought that this view opposed the sovereignty of God for it meant that God was determined by something external but it neglected the kind of creator that we are postulating: one who limits his sovereignty in order to create free creatures and respond to their actions.

Incomprehensible

We think of God as a mystery and our concepts as inadequate to grasp the fullness of his Being but we need not be influenced by the Kantian view that our minds structure reality resulting in agnosticism or follow Heidegger for whom Being in itself was unknowable. We do know in part, but what is revealed while suitable to human limitation is trustworthy. But it must be critically assessed for consistency and coherence with what we know about God's character. When an assassin tells us that God ordered him to kill Yitzhak Rabin, the Prime Minister of Israel, we know that whatever experience he had, it was not given him by God. The ineffability of God means otherness or distinction from us, majesty, thoughts which are higher than ours, justice, righteousness and holiness. It is clear from the visions of God which appear in the religious traditions that God does reveal these perfections to us though modified in accordance with our human understanding.

Omnipresence

God is everywhere and pervading all things not as a Being in space, for that could not contain him, but as Spirit. He is the creator of spacetime and is the ground and condition of both. When we say he is everywhere we mean that he makes his working felt everywhere. He pervades the whole without being included in the order of things which coexist in space.

In conclusion we note that the models mentioned in this chapter and the last will be helpful when constructing our own in the final chapter. Before doing this we need to ask about the existence of God, something which the scriptures of the various traditions take for granted.

10 Does God Exist?

If from the indubitable fact that the world exists, someone wants to infer a cause of this existence, his inference does not contradict our scientific knowledge at any point. No scientist has at his disposal even a single argument or any kind of fact with which he could oppose such an assumption. This is true, even if the cause – and how could it be otherwise – obviously has to be sought outside this three-dimensional world of ours.

Heisenberg

The various models of God that we have considered do not exclude one another and we will draw elements from them in understanding what God is like in our final chapter. But they do agree that God exists not as another object in the world but as the ground of all. He is an objective Being who manifests himself in the world yet is transcendent. The scriptures assume his existence and it was left to the various philosophers and theologians of each faith to argue for it. We will consider some of these arguments in this chapter. Philosophers, sociologists and psychologists have seen God as a projection of man (Feuerbach), the opium of the people (Marx), the illusion of those who have remained infantile (Freud), and the symbol of society (Durkheim). Such scepticism is not a particular symptom of modernity for Gibbon pointed out that the various modes of religion which prevailed in the Roman world were all considered by the people as equally true; by the philosopher as equally false; and by the magistrate as equally useful.

WHY IS THERE SOMETHING RATHER THAN NOTHING?

The answer of the theist is a God behind the something, that is, the universe. We ask when something occurs what was the cause? Hence the argument: every event has a cause, the universe is an event, therefore the universe has a cause. The Muslim, Jewish and Christian theologians, being influenced by

Aristotle, developed such arguments for the existence of God. In Islam appear two versions of this cosmological argument, that from temporal regress and from contingency or dependency. There were two schools: the Kalam (doctrine) and the Falsafa (philosophy). Their approach differed from the Christian in the denial of secondary causes for everything occurs by the direct action of Allah. But there is agreement in that Allah creates the world ex nihilo, it is temporal and contingent, and Allah sustains. The Arabic philosophers believed in God as the efficient Cause or as the rationale with the latter gaining prominence. Some of them preferred emanation rather than creation.

The Jewish philosopher, Saadia ben Joseph (882–942 AD), based his argument on the finitude of time but Maimonides (1135–1204 AD) was willing to accept that it might be eternal for it would still depend on God's will. He preferred a more personal image of God than a First Cause and opted for the Supreme Craftsman as in Plato's Timaeus. He developed four arguments for the existence of God which Thomas Aquinas followed, shown by the Five Ways using often identical phraseology.[1] The argument moves through a sequence of causes and effects to a First Cause or Necessary Being: one that contains within itself the reason for its own existence. But we are moving from a series of finite causes back to an infinite First Cause: a leap from the finite to the infinite, hence the argument is defective. How can a series in time give us the infinite? But what if such a First Cause does not mean priority but a vast superiority that is beyond our conceiving? When we use the term 'first' we think of 'before', that is, temporal. I decide to leave the room and get up from the desk and open the door. Decision and rising come before and are the cause of the door opening and have temporal priority producing the effect of the door opening, but if I am standing beside the door and just open it, cause and effect are simultaneous. Hence we do not think of the First Cause as in the time sequence, and in any case time itself was created with the universe. In speaking of a First Cause we are thinking not of the Big Bang, that is, physical cause, which brought the universe into being but the reason why we have one. It is an ultimate explanation not a Cause and the answer to Hawking's question: why there is a universe, not how there is one.

David Hume was sceptical of the conception of cause and effect. He argued that it is only the repetition of events in association which makes the imagination connect them and describe one as causing the other. It is a subjective association of ideas: a sequence. Hume's scepticism was criticised by Bertrand Russell who accused him of destroying all rational thought.[2] Hume believed that ideas flow from the senses which cannot establish supernatural causes and Feuerbach in similar vein concluded that we cannot think of a nature higher than our own. Hume's arguments resemble that of Ramanuja who contended that you cannot infer from how things are in the world to the Cause of it. In the Hindu cyclic view of the world there would be separate creations and even if we were successful in inferring a resemblance we would only have a superhuman person. He and David Hume think that we cannot move from the human to the divine but we have argued that human analogies used by religions are useful provided that we recognise both likeness and unlikeness in understanding God.

Instead of thinking of a series of causes we might consider a row of boxes placed each on top of the other. Thus the foundation box will need to be something special, particularly as the number of boxes increases. By analogy God is the foundation of all that follows, distinct and omnipotent, ensuring that the universe is sustained. Such a ground is not another cause in the temporal sequence but supreme in the ontological order. It is necessary, that is, self-caused, and unique.

WHAT ABOUT DESIGN?

Bertrand Russell insisted that the universe is just there and no further discussion is possible. The reality of the problem is denied by refusing to enter into argument about it, but as Coplestone said in debate with him: 'If one refuses even to sit down at the chess board and make a move, one cannot, of course, be checkmated.'[3] It does seem odd that we demand reasons for everything yet accept that the universe is just there. Science informs us that the universe will end in the Big Crunch so it is finite and cannot contain an explanation within itself for its existence. When pressed Russell said that the universe was due to an accident or coincidence but we noted that the order and precise tuning opposed this view. A physicist writes:

> If the big bang was just a random event, the probability seems overwhelming (a colossal understatement) that the emerging cosmic material would be in thermodynamic equilibrium at maxim entropy with zero order. As this was clearly not the case, it appears hard to escape the conclusion that the actual state of the universe had been 'chosen' or selected somehow from the huge number of available states, all but an infinitesimal fraction of which are totally disordered. And if such a exceedingly improbable initial state was selected, there surely had to be a selector designer to 'choose it.'[4]

Hawking also notes the precise timing for he says that if the conditions and constants in the beginning of the universe and the rate of expansion one second after the big bang had been smaller by even one part in a hundred thousand million million, the universe would have recollapsed before it even reached its present size.[5]

In Indian philosophy we noted the concept of karma which is employed as an explanation of a major objection to design, that is, evil. It would require Brahman to administer but it is the reason for the suffering caused by human wrong doing. Karma is considered to be a natural law just like gravitation. But why does Brahman permit karma to exist? Because he wanted souls to work out their karma and this would involve suffering as a discipline. In that sense the law is part of Brahman's nature and inequalities are explained as the result of sin in a previous life. The doctrine rests on the validity of the transmigration of eternal souls, 'the wheel of life' (samsara), otherwise the question arises: when did the responsibility for what one does first occur? With karma there seems to be a mixture of predestination and free-will. The Indian tradition does not speak of God as judge who will acquit or condemn at the end but karma might be seen as similar to the principle that what we sow we will reap.

It is somewhat ironic that David Hume and Immanuel Kant who were critical of all the traditional proofs for the existence of God paid respect to the argument from design. Hume with his usual acuteness, said that an organising principle responsible for patterns in nature might be within organisms not external to them. Today it is this which is stressed. God has ordained nature to make itself through a long history of evolution, that is, design is built-in. Purpose implies a mind at work and even those who think that the accidents of physics and astronomy produced a hospitable place for humanity and do not claim that the architecture of the universe proves the existence of God, go on to admit that 'the architecture of the universe is consistent with the hypothesis that mind plays an essential role in its functioning'.[6]

Those scientists who say that science proves that there is no divine purpose are making a metaphysical not a scientific claim. It is materialism. Scientists say that teleological causes are not useful in their work but that does not eliminate an overall purpose. The reason or cause of an individual thing is not the same as enquiring about the cause of a totality. To insist that there is no overall purpose is a leap of faith for it would require an exhaustive and unlimited knowledge of the whole universe. Darwin, we recall, said that he could not think of the totality of the world as the result of chance. His difficulty was seeing each separate thing as the result of design.

Instead of a causal chain we think of an explanatory one. And, instead of a predetermined blueprint, we might consider the world as a vast experiment with God improvising in an open-ended process.[7] Chance does point to this but it does not rule out order, for from chaos, order can arise. The interplay of chance and law allows new forms to emerge in the evolutionary process and such laws express God's purpose in the world. From the basic molecular structures of matter to the organisation of our bodies distinctive patterns and complexity emerge showing order and design. If, however, there are many universes it may be that our universe only appears to be designed. It is by a lucky chance that a universe having that form is one where life arises. This proposal introduces a complexity, that is, many universes, that opposes Occam's razor which calls for a simple explanation. It is a metaphysical not a scientific explanation since we only have the experience of one actual universe. Other universes, if they exist, are not accessible to us, we cannot observe from ours or devise some experiment to discover them. Once again we are asked to believe in an accident or lucky chance as opposed to the theistic view of a divine purpose. The choice may be compared to execution by a firing squad. As the firing ends I am surprised to discover that I am still alive. Have they all missed? A lucky chance or have they done it on purpose![8]

There is some advantage in thinking not only from design but to it. Evolution places the stress on development and the theist sees things proceeding to a climax or goal. God is the Final Cause and lures us towards the ideals of goodness and love. These are higher influences unlike physical causes, which again is not odd because in physics there is a parallel: an electron in a probability distribution does not have a definite position so that no physical force at all is needed to influence the outcome. The most significant stage in the evolutionary process has been seen as the change from instinctive behaviour to that which was a conscious response to value.[9] Unlike animals we aim at truth, beauty and goodness, and it would appear that this is the goal of evolution for it is manifest in its crown: mankind.

THE IDEA OF GOD

The two arguments which we have considered spring from the experience of the world but the ontological stems from an idea. It is therefore a priori rather than a posteriori being independent of experience. Anselm (1033–1109), Archbishop of Canterbury, had to contend with King Henry's desire to rob the church of its revenue but he found time to define God as 'that than which nothing greater can be conceived'.[10] To exist in reality is greater than to exist in thought alone, hence God must exist. How can I postulate every attribute of perfection to God and yet exclude existence? But I can think of the most perfect island without it existing. Such an island, said Anselm, would be an object in the world and would owe its existence to natural causes but God is the ground of all and self-existent. Descartes supported Anselm's view: God has existence just as a triangle has three angles. From the definition of a triangle we can tell that the sum of its interior angles equals 180 degrees. A triangle without its properties would not be a triangle and God without existence would not be God. We could not have the idea of imperfection without the conception of perfection.

Immanuel Kant was not impressed. He agreed that if there is a triangle it must have three angles but the question is if! Kant denied that existence is a predicate or possession: I can think of £100 but that does not mean that I possess the money. Kant's point is that an idea is a concept, not a reality corresponding to some object and the properties of a thing do not include existence. Imagine that you are interviewing candidates for a job and you study carefully the properties that they possess: appearance, education, good health, reliability, but existence is not one of them!

Norman Malcolm thinks that Kant's criticism refers to contingent things but God is a necessary being, greater than anything we can think about. On the basis of Wittgenstein he points out that there are different language games and in the religious language game, God is a necessary being. He does agree that when we speak of necessary propositions: all bachelors are unmarried men, it simply represents the use of words or the relation of concepts but that does not make the existence of God null and void. In fact when we use the word, 'God', it implies necessary existence as Kant admits. How then can he go on and say that he does not exist? You cannot have it both ways! If I am asked what a square is I will say it necessarily has four sides, and it would be irrational to say in the next breath that it does not have four sides! Malcolm believes that once we know the meaning of the word 'God' we will see that he necessarily exists and we cannot then go on and query such existence.[11] Malcolm is treating God as a unique case for he has all the perfections which must include existence. His argument has been criticised but it is helpful in pointing out that there are different kinds of discourse and statements which have various evidence for their support. Things exist in different ways and at various levels.

WHAT ABOUT MORALITY?

If Kant bowed God out of the front door of the house in his *Critique of Pure Reason*, 1781, he let him in by the back door in the *Fundamental Principles of Metaphysics and Morals*, 1785, by postulating the moral argument for his existence. Two things caused Kant to wonder, 'the starry heavens above and the moral law within'. It is the moral law which insists that I ought to do my duty, impinging upon me with the force of a categorical imperative. The 'ought' implies 'can' for it would be absurd to feel obliged to do an action if I did not have the freedom to do it. It would be like telling a beggar to donate sums of money to charity! The possession of free-will distinguishes the human being from nature where things are casually determined. We must do our duty but it is often unrewarded in this life for we see the good person suffering and the evil one prospering. It is also clear that the ideal of goodness is never reached, hence in a just order of things there must be another world or state in which virtue is rewarded and the ideal reached. And there must be a Lawgiver to apportion such happiness.

Such an innate sense of 'oughtness' has been challenged and explained by the nurture we receive from family and society. Darwin and Wallace disagreed about the source of values but if there are genes that cause disability and criminality could there not be religious and moral ones? Many people have been successful and triumphed over their initial environment whereas some favoured by a good nurture have failed to take advantage of it. With Kant there is the phenomenal world of appearances, perceived by the senses and the noumenal that lies behind it. The theoretical reason cannot penetrate the noumenal hence his rejection of the arguments for the existence of God but he thinks that our sense of obligation or duty comes from the noumenal. Kant never really explains his two worlds and faith and reason cannot be separated in this way. He postulates a just God who is a celestial paymaster reflecting one facet of the religious traditions but it is only one aspect of their model. Kant believed that rational creatures would recognise the values and aim at them but our will is weak and inclined to evil rather than good: 'the good that I would I do not and the evil that I would not do that I do' (Rom.7.19). Hence the need for the grace of God.

Our world does have a moral purpose but the achieving of goodness involves suffering as the Indian philosophy teaches. Life emerges from death in nature and suffering can produce beauty: it was the deaf Beethoven who composed the Nineth Symphony and the blind Milton who wrote *Paradise Lost.*

MIRACLES

Miracle was used traditionally as an argument for the existence of God. It has been defined as a violation of nature or as the transformation of physical objects by the power of God. The world that we have described in Chapter 5 does appear more open for the interaction of God than the Newtonian mechanical particles determined by fixed laws. It is governed by statistical laws which do not determine occurrences of small events only proportions in the larger classes of events. Therefore it is difficult to know how nuclear particles behave under certain conditions and observations and, if there is probability not certainty, it is impossible to regard the existing state of science as finally legislative of what is going to occur in nature. Physical science is credible but not certain and the human sciences show that it is more problematical to predict the laws of our complex human nature.

Laws of nature are based upon observation, experiment and experience. On the basis of experience we expect that when A appears B will follow but it is not always the case. Better weather can normally be expected in England in June rather than December but then the unusual happens in a certain year: a mild December and a bitterly cold June. The uniformity has been broken and the meteorologist tries to explain but the weather is a sensitive system. Relying on past observation she infers that it is unlikely to happen again but since she cannot observe the future there is no certainty. On the basis of observation and experience all we can say is that A is normally followed by B but there may be exceptions to the regularity.

Science is based upon experiments which take place under controlled conditions and they are repeatable. To prove or disprove something experiments are performed again and again and it only requires one exception to falsify a law. But miracles are not repeatable: they are particular and peculiar events occurring in human situations, '… they are not small scale laws and consequently, they do not destroy large scale laws'. A miracle is unique because it is unrepeatable but it is not sufficient reason to abandon natural law for to do that it would have to be an experimentally repeatable exception: '... the miracle does not fall into this category otherwise it would itself be a new small scale law ...'.[12] Miracle has the peculiar power of violating but not destroying a law of nature. Of course whether or not a miracle occurs depends on the evidence: a wise man, as David Hume said, proportions his belief to that factor. It is different from saying that miracles do not happen because of an a priori ruling that they cannot happen, but Hume fails to remember this. The evidence for each miracle must be carefully sifted and reasons given for the supernatural or mythical or naturalistic explanation.

Healing miracles are not as suspect as natural ones but miracles need to be related to a wider conceptual scheme that makes sense of their occurrence. They are not magical tricks or random events but occur in a sequence and relate to the self-disclosure of God. The world is disordered by sin, disease and death, and the miracles of Jesus bring the new order of healing and life. They were not done to impress or to create belief in him but to meet human need. In this interpretation miracles are a part of revelation not a proof. But other explanations can be postulated: reports of miracles are mistaken or can be given a naturalistic explanation or God acts alongside other causes, and so on. But even a naturalistic explanation of the miracles of Jesus credits him with amazing insight and compassion. Many people prefer naturalistic explanations because they see a miracle as Hume did: a violation of natural law. But such laws are formulated by a process of induction which is not foolproof and there is no need to think of God breaking into the natural order but present in and interacting with it. Laws can be falsified and reinterpreted: Einstein's view of gravity is not that of Newton. Laws are descriptive, relative and provisional, not absolute certainties. They do not make things happen but describe how they occur.

Hume argued that the witnesses to miracles were primitive people who believed in them but we cannot. But in the religious traditions there are intelligent and educated witnesses. We grant that the mass of people are often gullible, for example, in the 16th century there were all kinds of belief in magic, superstition, relics and astrology. Today, despite our sophistication, such beliefs persist as tales of the abnormal and supernatural show. But we cannot write off the New Testament records on this basis. We have seen the intellectual and scientific advance of the Greeks in Chapter 1 of this book. It is no wonder that their philosophers laughed at Paul when he spoke of resurrection (Acts 17.32) and that the crowd on the day of Pentecost thought the disciples were drunk (Acts 2.13). Thomas was a doubter, and as we mentioned, earned the approval of Richard Dawkins. The enemies of Jesus were so sceptical that they put his power down to the prince of demons: Beelzebub (Mt.12.24). It is clear that the world view was different but sceptics of divine power were present then as they are now.

Hume also contended that miracles occurred in many religions, but why should they not happen if God has revealed himself everywhere? The stronger argument concentrates on the breaking of the regularity of natural law, thus even if there is evidence to the contrary he is not willing to move from the assumption. But such a view opposes Hume's own argument about causation. There is no necessity, according to Hume, that what has always happened must happen and what happens today may happen tomorrow. On this basis, Augustine's definition of miracle has force: 'Miracle is not contrary to nature but what is known about nature'.

Miracles do not always create faith but those who have faith can create miracles: believers have sometimes managed to do what was thought impossible. Miracles need not be expected if faith is absent (Mk.6.5:6). A prior belief in God as a Being-in-action and the creator of a novel world expects the extraordinary. Hume did not have that belief and his philosophy led to scepticism about what we could know yet he is certain that miracles cannot happen! Perhaps our previous discussion of the relation of mind and brain might be of help. Mind emerges from matter and is connected with wholeness: a water molecule is not 'wet', yet when many of them come together they produce water, proteins and acids, and form a cell. A neuron in the brain is not conscious but with other neurons there is consciousness. The emergence is due to the pattern or the arrangement and once a whole is formed there is 'downward causation'.[13]

How does it apply to mind and body? I am typing a paragraph at the moment but decide it is time for coffee and rise to make it. While typing, one pattern of activity is going on in the brain, but my decision to make the coffee causes another pattern to form and physics cannot tell me why this should be. I have interrupted one flow of activity and replaced it by another. But the neuroscientist will say that having observed my brain for a long time he can see signs of tiredness and can predict the break for coffee. It is all a natural process, and it seems to eliminate free-will. But as Russell Stannard points out life is not like that for it makes no sense to say: 'I don't have to make up my mind when to stop reading (or typing); when the working out of the laws of nature in my brain dictates that I shall stop, then I shall stop'.[14] Without a positive decision I am not going to get that cup of coffee! A downward causation or influence is necessary. We think of God acting in this way with regard to the world. A miracle would not be an intervention or disruption but interaction or working within the causal laws to bring about an unexpected event. Our analogy about the coffee fails, however, in that it is a normal event but we can say that there are occasions when we do the unexpected and un-predictable. A miracle is a 'wonder' or 'a sign' which departs from the ordinary, otherwise it would not cause astonishment. The test of miracles appears to be their effects. If they are sen-sational and bizarre but do not accomplish any good they fail to meet the test of what religion says about the character of God. He is just, loving and compassionate, and therefore when a miracle results in righting wrongs, healing or relieving the oppressed, it deserves not to be dismissed by scepticism about them in general.

RELIGIOUS EXPERIENCE

Religion is natural to mankind and Sir Alister Hardy at the Religious Experience Research Centre, Westminster College Oxford, recorded that 50 per cent of people involved in a survey said that they had some kind of religious experience. Some experiences are extraordinary according to the religious traditions. Rudolf Otto (1869–1937) in *The Idea of the Holy*, argued that the centre of religious experience is a holy, awe-inspiring Being, a mysterium tremendum atque fascinans. It induces a feeling of guilt, makes us tremble, but fascinates ...

The anthropologist explains it by mana, magic and ritual: expressions of the hopes and fears of primitive mankind. But Otto was able to show that the sense of the 'wholly other' extended from early man to eminent scientists like Pascal, and writers and poets such as Coleridge and Ruskin.[15] The fear we experience of a tiger springs from the sense of danger but it does not have the uncanny nature that chills us when we look at a dead person. The numinous that Otto considers has a paralysing effect but also a thrill of excitement for it attracts, fascinates, and draws towards the ideal. He stressed the non-rational, beyond reason, not the irrational: a mysterium which transcends normal experience.

These experiences often occur at some particular crisis: a disgust with the commercialisation of the Church (Luther) or with one's spiritual experience (Wesley) or facing a battle (Arjuna) or ... . There are also crisis periods in science as Kuhn noted where a new paradigm is envisaged that cannot be settled by logic and experiment alone.[16] Religious experience need not be extraordinary but the ordinary seen in depth. The poet inspired by the beauty of nature experiences 'a presence that disturbs [him] with the joy of elevated thoughts', the psalmist reading the Torah (law) finds 'new life for the soul' (Ps.19.7), and the old countryman taking off his cap when he sees a sunset of unutterable beauty says softly: 'it is God'. Eligah found God not in the whirlwind or the fire but in the 'gentle whisper' (1 Kings 19.12).

There is a place for a study of the observable phenomena of religion – sacred objects, social institutions, actions, customs rites – for we need to know the actual facts about it. But to reduce it to this is not to do justice to the experience of God. Religions are living communities and involve the personal convictions of their members and cannot be classified as lifeless specimens in a museum. Arvind Sharma, a Hindu philosopher, points to a parallel with science. A schoolboy reads in his textbook that water is H_2O and conducts the experiment. It is a success and now he knows for himself that water is H_2O. He does not need faith in the text or the teacher any more; he knows it for himself. Even during the course of the experiment only provisional faith in the teacher and the text were required but he had to trust the method otherwise there would be no experiment.[17] The analogy breaks down when we think of the public nature of the experiment and the private nature of religious experience but it will have to agree with scripture and the experiences of the community of believers. J. S. Mill regarded utilitarianism not only as a philosophy but a religion in the best sense. He thought his ethics did not differ from that of Jesus of Nazareth. There was a crisis in his life (1826) which some understand as a mental breakdown and others as a crisis of faith. He was aware that the analytic method of philosophy had worn away his feelings which were the source of happiness. Attention, he realised, must be paid to the inner life which could be developed by the arts. There is a resemblance with Darwin who confessed in later life that he had lost his appreciation of music. William James (1842–1910), examined a wide variety of religious experience and did not neglect feelings. He concluded that it resulted in inner peace and love, not neurotic but a stimulus to good living. As a pragmatist, James held that there is truth in a belief if the results are good. The question was not: 'does God exist?' but 'what difference in conduct does belief in God make?'. But must truth not correspond with some object or fact as well as working in our experience? Is it not possible that belief could be an illusion even when it leads to good results? Some instrumentalists in science would agree with James: if a theory works, use it. But most scientists hold that a theory must correspond with the facts if we are to believe it. James would have replied that even if reason makes us doubt religious experiences it is better to believe than disbelieve since it helps us to live positively. Truth is not known by thinking about it but acting just as love is known by loving, feeling, and involvement.

Thomas Hobbes, however, doubted religious experience. He remarked that when a man says that God has spoken to him in a dream it means that he dreamed that God spoke to him. Hobbes could have pointed to extremes in this connection. Among the Sufi mystics some declared their unity with Allah to such an extent that Hallaj said that he was the Truth. Then there are those who engage in mysticism, which does not necessarily mean that they experience God, for example, the Buddha. Mysticism has appealed to many great scientists: Einstein, Pauli, Schrödinger, Heisenberg, Eddington and Jeans. Ninian Smart is helpful on the difference between mysticism and prophetic experience. First, the mystic looks within, into his own soul and beyond. But the prophet has a vision exterior to himself of someone who is holy and grants illumination. Second, mysticism can occur, as in Theravada Buddhism and elsewhere, in a context where there is no concept of a creator and does not result in any kind of relationship with him. Third, the language of the prophet about his experience is personal, whereas the mystic's is frequently impersonal. Fourth, the experience of the mystic leads to a stilling of activity whereas the prophetic is dynamic and active. But it may be asked: can the external and internal be so sharply defined? An external event produces an inner feeling and searching my feelings can make me think of something external.[18]

Sigmund Freud's (1856–1939) examination of religious experience, as is well known, resulted in a negative verdict but Carl Jung (1875–1961) was much more positive.[19] Karl Popper finds fault with Freud and Adler, because they were always seeking confirmation of their theories, instead of trying to falsify them. He points to Einstein's theories as being open to falsification and willing to take the risk of being proved wrong but with Freud and Adler almost any case would prove them right.[20]

Sociologists as well as psychologists have been interested in religious experience. Early social evolutionary theories explained man's experience of God as derived from feelings of awe invoked by mountains, rocks and waterfalls and the wild landscapes that he encountered. But Emile Durkheim (1858–1917) equated God with society and saw religion as a unifying force, a product of the social consciousness giving a sacred backing to the laws. Every society found it useful in confirming its hierarchy. Hence it was not animism or mana or magic that produced the idea of God but society itself. Religion can protect the status quo: 'The rich man in his castle, The poor man at his gate, God made them high and lowly, And ordered their estate.' But such a view does not agree with the Magnificat (Lk.1:46:56) which envisages a moral, social and economic revolution. Or with the involvement of religion in liberation movements throughout the modern world. If a state is unjust religion must oppose otherwise it deserves Marx's stinging remark: 'the opium of the people'. Religious experience points to something beyond

society: 'signals of transcendence'. Peter Berger contended
that our desire for order and purpose, our engaging in play
which lifts us above the dull routine, our hope for justice that
is not satisfied with its poor imitation on earth, our sense of
humour which enables us to rise above the tragedy of life,
point to a perspective that transcends.[21] In similar vein Max
Weber (1864–1920) saw the impact of charismatic personalities on societies and that Calvinism could bring change with
its stress on hard work, thrift and the avoidance of frivolous
pleasure. Religion could mould a society or produce a new
one: Islam is a classic example.

The Buddha was reluctant to tell others about his experience just as Muhammad and the Jewish prophets were not
only unwilling to speak the word of the Lord but were
horrified at the things they were required to do. If what these
men experienced was an hallucination it must have had
tremendous power for it went against their ideas and exposed
them to despair, frustration and death. How can religious
experience be a projection of our wishes if it calls us to go
against them? It might be argued that the mind is deranged
and indeed Muhammad thought he had gone mad but was reassured by his wife. Sometimes it results in strange behaviour
but generally the remainder of the life shows a dedication to
the service of God and humanity.

A. J. Ayer, who placed such stress on empiricism, records a
religious experience.[22] In 1988, after a bout of pneumonia, he
was rushed into hospital, where his heart stopped for four
minutes. He was not dead, for death is connected with the
brain stem and not with the stoppage of the heart. In an
article published in the *National Review* of October 1988 he
recorded that during these four minutes he saw a red light
which he thought was responsible for the government of the
universe. He observed two ministers who were in charge of
space but space itself appeared out of joint. It was impossible
to extinguish the light since the laws of nature did not seem to
be functioning. When Ayer died a year later his colleagues said
that the experience was hallucinatory but writing about it Ayer
reflects on the question of an immortal soul or the resurrection of the body. He points to the latter as being the Christian
view and thinks that a reunion of the same atoms with more
than a strong physical resemblance, that is, similarity of behav-

iour might be possible. Atheists, he says, have believed in immortality and since he did not experience God he concludes that it is possible to accept the former without the latter belief. He thinks that his previous conviction that there is no life after death has now been 'slightly weakened', but he remains an atheist.[23]

Perhaps the tests of religious experience resemble more the artistic: many aesthetic experiences are not naturally shared by a majority and their development requires attending music or artistic appreciation classes. There is no universally adopted testing system for works of art and no sophisticated predication for statements about art. Just like religious experience the aesthetic cannot be assessed in a quantitative way by measurement but through the study of the great masters. Thinking, listening, absorption in the art form, inspiration and feeling are necessary in music and painting, as in meditation and prayer.

GOD EXISTS IN RELIGIOUS COMMUNITIES

Philosophers in the 20th century questioned the language of religion and morality. They argued that only empirical statements, verified by sense experience, and analytic statements which say what we intend words to mean, were possible. A religious or moral statement appeared to be neither of these and was considered to be an expression of feelings. But it is now generally accepted that this was a limited view and ruled out its own verification principle.[24]

After Wittgenstein the slogan concerning language became: 'Don't look for the meaning, look for the use'. He compared language to a bag of carpenter's tools, each with its own particular function or with a range of games each with its own rules and criteria. Language is social and cannot be uprooted from the life of the community in which it originated. Thus a stranger in a foreign country needs not only to master the rules of the language but study the culture, customs and traditions, if he wants to understand the people. Only when we see the use of language can we understand the meaning. The language of religion is intelligible within a Mosque, Temple, Church or Gurdwara, and is subject to internal rules. It does not make religion a private affair for the community has rules to regulate the behaviour of its members and they will govern their conduct in the market place. Wittgenstein did not think that God was unverifiable or non-existent but inexpressible.

The non-realist, basing his thinking on Wittgenstein, insists that God exists in the context of the life and language of religious communities. Statements are true because they fit a particular form of life. D. Z. Phillips insists that language gets its meaning from its context and religious language describes a way of life. God cannot be defined for he is inexpressible and if we try and capture him in doctrinal statements he becomes one existent among others and contingent like us. God cannot be inferred from the world but known by his role in the worship of a community. The model is true or false in that context. He refers to Wittgenstein's 'picture' theory of language, and points out that the religious believer's image of the world differs from the unbeliever because of preference and commitment. The view that God is the explanation of the universe is set aside since it would mean that he was in competition with other explanations. Phillips comments ironically on those who believe in the probability of God's existence. They have to change the creed and recite: 'I believe it is highly probable that there is an almighty God, maker of heaven and earth'!

Another non-realist is Don Cupitt who denies God as a transcendent reality. He is influenced by Buddhism and the concept of emptiness or nothingness. He then produces the curious argument that in order to save us God must make us nothing and to make us nothing God must himself be nothingness. What has gone wrong, he thinks, in Christianity is that it has portrayed God as a powerful Ego, self-existent and omnipotent, the Great I Am. It is incompatible with the Buddhist stress on the illusory nature of the ego and with what the religious traditions teach about our becoming less egoistic and more selfless. Cupitt will have little to do with metaphysics so he denies an objective God and immortality. We can go on worshipping in our various communities but worship should be directed towards the ideals and values which we hold, hence God is the personification of our values. Since he believes that language is unable to grasp certainties we deal with appearances not realities.[25]

There has been much debate about Wittgenstein's 'language games': language used by various communities, and whether or not he meant that it separated them and that they could not judge one another. With him the role of philosophy is descriptive but it also must be critical. Such criticism shows that all frames of references or views of the world and God are not equally true. Karl Popper opposed Wittgenstein, arguing that we not only criticise communities but break out of them. It is the realistic view. Of course we need to know how language is being used and the coherence of the whole system which would be one test of its truth. But the non-realist says that language is relative to the culture in which it is used and we must not look for correspondence to reality. It fits the beliefs of a society at a particular time but later it is shown to be incorrect. At one time people believed that the earth was flat in accordance with their world view but it is not within our frame of reference. In such a changing world we never have reality.

But faiths that isolate themselves from what is happening in the world around them become religious ghettos. Religious traditions change in their expression of the faith in different historical and social contexts but a test is how a contemporary communication is continuous with the beliefs of its founder. Muhammad, Jesus, Nanak and the prophets of Israel believed in the existence of God in some objective way and would hardly have recognised equating him with our values. The fact that revelation comes through the human does not mean that it is simply the expression of the human for it can oppose the values of community, our natural instincts and beliefs. It is true that one form of life cannot judge another by its own terms but truth is universal. Philosophy will critically examine all statements whether religious or not and there is no special sphere in which it is not allowed to work. Cupitt and Phillips want to make religious statements expressive or exhortative and not referring to some external Reality. Cupitt, stressing the inability of language to grasp God, argues that no certainties about him are available but if he believes such a statement, labelled perspectivism, to be true then he recognises a certainty! On the other hand if it is merely a point of view why should we take it seriously?[26]

In conclusion it is clear that the arguments for the existence of God are not proofs but it may be claimed that in a cumulative way they point to probability. Phillips is not happy about this but we live in a world that has few certainties. We would also contend that it is more reasonable to believe in God, for what is the alternative? Is the world an accident with no meaning or purpose and no life after death? Even A. J. Ayer modified his denial of the latter. Wittgenstein said: I know nothing of religion but there is surely something right in the concept of a God and of an after-life.[27] Dostoevsky had made an impact upon him and he had a religious conversion which induced him to give up his possessions and devote himself to inner goals.

The agnostic appears to want more reasons for believing than not to believe. But is this not too stringent a demand? Suppose I gather a considerable amount of evidence that Smith exists and the same amount that indicates that he does not. What am I to believe? Am I too much of an optimist to give Smith's existence the benefit of the doubt? And we remember that we are talking about Smith and God and they exist in different ways since God is a unique Subject as the arguments insist. Should we not because of the limitation of our knowledge and experience of such existence require less justification? Some argue that God should give more direct evidence of his existence instead of 'the whisper of his ways' but, as we noted, even if direct revelation was possible it would compel belief and be inconsistent with our autonomy. In science proof has often been delayed: Einstein had to wait for the eclipse of the sun to demonstrate that gravity bent light, and in religion it is contended that there will be eschatological verification at the consummation of all things.

We now turn in the final chapter to consider a model of what God may be like.

11 What Kind of God Model?

It seems to me that if the word 'God' is to be of any use, it should be taken to mean an interested God, a creator and lawgiver who has established not only the laws of nature and the universe but also standards of good and evil, some personality that is concerned with our actions, something in short that it is appropriate for us to worship. This is the God that has mattered to men and women throughout history. Scientists and others sometimes use the word 'God' to mean something so abstract and unengaged that he is hardly to be distinguished from the laws of nature ... but it seems to me that it makes the concept of God not so much wrong as unimportant.

Stephen Weinberg

In this final chapter we look at some of the insights that we have gleaned about the world, mankind and God. The religions differ about creation, soul, incarnation, nature of God and so on, so there is not a common essence. But they agree, apart from Theravada Buddhism, that there is a spiritual presence in the world and we are related to it. The world emanates or has been created by God and there is agreement that its values delude and cloud our understanding of beauty, truth and goodness. In most of the religions there is a theistic strain which we prefer in putting forward a model of what God is like.

From Hinduism we take the insight about levels of truth in connection with the scriptures. One cause of conflict between religion and science was the literal interpretation of scripture. But if we apply Sankara's levels to it then there is the recognition of the allegorical, moral and analogical. The rabbis of Judaism and the Christian Fathers concur that scripture uses signs and symbols to get beyond the literal. The first is easy to understand but the latter tries to express the inner meaning of outer events by transcending the material. What is observed at the lower level of the literal becomes the vehicle for the symbolic meaning. Both religion and science go beyond the observable in constructing their theories so an understanding of the symbolic nature of what is being said is essential if conflict is to be avoided. But it is necessary not to neglect historic elements in a religion in favour of the symbolic otherwise we can arrive at Paul Tillich's God above God. It is significant that John Robinson who followed Tillich in his writings draws back at this point and says that this is heady language, dangerous language, and it has a long theological history strewn with many warnings: 'Those who have spoken about going above the God of theism have usually sunk below it; and those who have aimed at a supra-personal Deity have ended more often than not with one that is less than fully personal.'[1] Tillich's God resembles Sankara's Brahman who exists on a different level from its manifestations in the world but, since we have made revelation primary, this is not acceptable.

From Hinduism and Sikhism we take the principle of identity-in-difference which can be applied to humanity and the sub-human. In Chapter 6 we saw that there was continuity yet discontinuity between the animals and man. Human beings are biological entities but much more, for a stream rises higher than its source. Man is a multi-levelled organism and his distinctive characteristics emerge at the higher levels. Cultural evolution is now more significant than biological with its transmission of information through memory, language, tradition, education and social institutions rather than by genes. In this way the human species has outstripped any other. But the materialist concentrates on the identity or continuity and fails to recognise the difference and thus presents an incomplete explanation of what we are.

Religion accepts that we are in continuity with the sub-human for we are from the 'dust of the ground' (Gen.2.7), but distinct because related to God. The difference between God and mankind is seen in our imperfection stemming from selfish desires that lead to suffering. Christianity postulates original sin which is confirmed in the genetic defect of the 'selfish gene'. There must have been a time when human beings, like other animals, had no concept of right and wrong so its acquisition meant that values enter in. This occurred at the human level and can be seen as the goal of evolution, that is, the moral purpose intended by the creator. But our imperfection expresses itself not only in our lack of concern for others but ignorance of the Cosmic Self within us or disobedience to God's commands. The theistic strain in the traditions teach that we require the help of God to experience forgiveness, release, happiness and the ultimate attainment of nirvana, heaven and paradise. Theravada Buddhism, however, calls for self-effort and Richard Dawkins says that we have the power to fight against our genes. But theism appears in Hinduism when Krishna says:

> Those who cast off all their works on me, solely intent on me, and meditate on me in spiritual exercise, leaving no room for others, (and so really) do me honour, these will I lift up on high out of the ocean of recurring death, and that right soon, for their thoughts are fixed on me. On me alone let your mind dwell, stir up your soul to enter me; thenceforth in the very truth in Me you will find your home. (Gita 12.6–8)

Grace is prominent in the religions. It is clear in Christianity, Sikhism and in Mahayana Buddhism where the devotees of Amida show a faith in the saving powers of a supernatural being. In Islam, Allah is compassionate but it is in the Shi'ah with its faith in Husain who died so tragically in the massacre of Karbala that there is the theme of redemption. They not only commemorate his sacrifice but plead its merit and in that sense he is considered a saviour.[2] Hence the assertion that there is no redeemer in Islam needs qualification.

While we accept that there can be meditation without God it is possible that prayer may overlap with it for a lot may go on in our mental processes of which we are unaware. The former lays stress on human activity whereas the latter speaks of God's initiative. But there are signs of cooperation in scripture: Paul wrote: '… Work out your own salvation with fear and trembling for it is God who works within you to will and act according to his good purpose' (Phil.2.13). It is the paradox of grace enunciated by D. M. Baillie: we know that God is doing it but we feel that we are doing it![3] A deterministic view of the sovereignty of God reduces the human to a state of passivity. If it is objected that faith is the gift of God we can reply that gifts can be refused. The debate divided the Church in the 16th century and the followers of Ramanuja in Hinduism but in Sikhism despite the stress on the sovereignty of God the recipient of grace is active.

We can think of faith as insight, 'the seeing of the invisible' (Heb.11.1), that is, seeing into the reality of things or beyond the material. This is what our discussion of symbols indicated. The Hindu is trying to do this and speaks of jnana: wisdom or knowledge, and the Buddhist of manas which is the sixth sense that coordinates the perceptions of the other five to give a complete representation of the object perceived. It is concerned with the relationship of subject and object.[4] The Hindu is meditating in the belief that he can achieve union with Brahman who is within him so he may be unaware of help from that quarter. When Peter in the Gospels made his declaration of faith in Christ he did not realise that it had been revealed to him until Jesus told him. Perhaps there is something in the suggestion regarding the Buddha that he was unaware of the presence of God in his experience due to his distrust of the multiplicity of Hindu gods. There are influences upon our lives that we are unaware of at the time and there is much in our unconscious minds that is hidden from us. We saw in Judaism that the Spirit of God was everywhere not only seeking to make God known but imparting technical gifts. If this is so then the insights of scientists could be given by the Spirit. In the creation of theoretical models thought experiments, imagination and intuition play a vital role. Both in religion and science insight results in creativity that departs from the normal routine.

With regard to the world, the Indian cosmology with its enormous time space and spatial scale is much closer to the immensity of the universe that science describes. It is immense with one hundred thousand million stars and galaxies reflecting its mystery and design, for if gravity had been stronger or weaker or the nuclear processes faster or slower it would not exist and there would have been no material for life to begin. The laws of nature united by some random process is an inadequate explanation as is the speculative view of many universes. The Newtonian machine model has been replaced by a web of relationships stretching down to the sub-atomic level so that one thing cannot be understood without reference to another and accords with the interdependence that Buddhism describes. God gives the gift of order and self-maintainance to the universe not once for all but continually as is observed in the initial conditions, the complexity of internal structure and the range of open possibilities. There is a uniqueness in these earthly factors because on other planets chemical creativity never advanced beyond simple compounds. The earth because of its own internal dynamics and its position in the solar system permitted chemical creativity and produced the first living cell.[5] God can be seen as the maintainer and sustainer drawing out the possibilities which are inherent.

Relation and influence replace rigid cause and effect. It was once thought that anything which influences the position of an object would have to be some sort of physical force but in quantum theory, an electron in a probability distribution does not have a definite position so no force is required to alter the outcome. We noted that natural selection was not the only element that could explain the evolutionary progress. By itself it cannot explain the initial mutations that generate more accurate information processing systems or that the environment would be friendly. It is contended that it is too precarious a situation to be left to itself by God, who intends rational beings to come into existence. The influence of God can be seen in 'a top-down' way, that is, the whole influencing the parts. Chance operates at the level of mutation in the DNA and does not prevent regular trends and the displaying of inbuilt propensities at the higher levels of organisms, population and eco-systems. It can be understood in an experimental sense at the micro-level so that the potential forms of organisations of matter might be explored.[6] God can be seen as the experimenter responding and initiating new courses of action and attracting the development of positive potentialities. It is cooperation, not interference, and unlike physical causality which is quantifiable and detectable. We know that influence plays a large role in our lives but we cannot measure it as we do physical things. How can ideas, feelings and personal relationships be quantified?

Concerning God we take from Hinduism the model of Being and Becoming. The Gita is intent on preserving the balance. With reference to Being and Becoming we are told,

'It is not Being nor is it not-Being' (13.12). Hindu thought
refuses to consider the either/or but advocates the both/and.
The Gita sets out to bring the two together and in so doing
realise the true spirit of the Upanishads whose description of
the Absolute is both a dynamic reality and yet eternally at rest:

Unmoving – One – swifter than thought –
The gods could not seize hold of It as it sped before [them];
Standing, It overtakes [all] others as they run;
In it the wind incites activity.
It moves. It moves not,
It is far, yet It is near;
It is within this whole universe,
And yet It is without it

(Isa Upanishad, 4–5)

A balance is struck between transcendence and immanence or
Being and Becoming. The Becoming, however, does not mean
that God is acquiring something in his interacting with the
world but actualising his character. It opposes the process
theology of an emerging God who is so affected by the world
that he appears to be struggling in it. Immutability reflects his
unchangeable character which remains in his interaction with
the world.

Spatial images either of height or depth are not to be taken
literally but metaphorically showing the uniqueness of God. In
our experience the self transcends relationships with others,
judges them, reaches for ideals, and asks about the meaning of
life. In experiencing God we cannot separate the external and
internal for an event will spark off reflection and how we are
related to it. God's becoming is described in a personal way by
A. R. Peacocke: 'It is of the nature of the personal not only to
be capable of bearing static predicates, referring to stabler
settled characteristics, but also of predicates of a dynamic
kind, since the flow of experience is quintessential to being
a person.'[7] The Trinity model stresses the personal but not
in the modern sense of a person for he is the eternal and
non-contingent source of personhood.

To influence the world God would need to have a close relation with it and the Hindu doctrine of emanation rather than
creation achieves this. In the Gita there is a sexual image of

creation with Krishna the incarnation of Vishnu casting his seed in 'Great Brahma', that is, material nature. Thus from the primal sexual act the whole universe comes into being (Gita 14.3). The process is evolutionary for the mind evolves from matter, then the ego which is the centre of personality, and finally the senses. There is a combination of goodness and evil with passion or desire and darkness or inactivity but also the good which causes the self to cling to wisdom and joy. A point of agreement can be seen with the Semitic view in that no material existed from which the world emerged and there is in the Hebrew Wisdom literature, wisdom as a woman, who is the creative agent of God (Prov.8;Wis.9). Peacocke writes: '... Mammalian females, at least, create within themselves and the growing embryo resides within the female body and this is a proper corrective to the masculine picture – it is an analogy of God creating the world within herself ... God creates a world that is, in principle and in origin, other than him/ herself but creates it, the world, within him/herself.'[8]

The mother image is a much closer relation than creatio ex nihilo but it is the Father which is primary in Christianity as the creed states: 'We believe in one God, the Father Almighty, Maker of heaven and earth'. Omnipotence is connected with the Father so we do not think of it in an abstract way of determining but permitting as a human father does. The close relation between the creator and us is illustrated in the use of the Hebrew word 'bara' which has a deeper significance than 'asa' which means to make. God breathes into man putting something of himself into him. In the Judaeo/Christian tradition the relation of parent/child is evident: Israel is a son who is disobedient, Jesus teaches his disciples to call God 'Father' and St Paul speaks of Christians as adopted sons of God. Islam, however, does not encourage the use of Father for Allah does not beget. But it is not a physical paternity which is meant, the use of the word is metaphorical for God is spirit.

There is also the belief in Israel that the relationship is an act of grace, of adoption, based on a free choice and taking effect in a covenant. The purpose of a covenant in the Bible is to link two parties who are not naturally united whereas there can be no question of a covenant between father and sons who are such by nature.[9] C. H. Dodd thought that there were only three places in the New Testament where mankind might be regarded as the offspring of God: Luke 3.38, Acts 17.28 and James 1.18. 'Father and son', as applied to God and man, mean his moral and personal protection and care. Of these the only clear case is that of Acts which is a quotation from Aratus who was a Stoic. Paul may have used the quotation to get a hearing from the Greeks for the whole tenor of his teaching is that we become sons and daughters of God by adoption and grace. The metaphor implies the restoration of a relationship which has occurred through Christ who is the Son of God by nature. Nevertheless, it means a close and personal concern for us and opposes any attempt to make theism look like deism.

We prefer creation for it preserves the distinction between God and us but if the creator breathed into us (Gen.2.7), set eternity in the human heart (Eccles.3.11), and there is a light that lighteth every man that comes into the world (Jo.1.9), then there is some affinity with emanation. But how is creatio ex nihilo related to science? Three reasons are given for the origin of the universe in the Big Bang: the space between the galaxies is expanding, the existence of the afterglow of heat radiation, and the abundance of chemicals stemming from the dense phase following it. Space and time are not the background but part of the universe and came into existence with it. Hence 'before' has no meaning which affects arguments moving from cause and effect since they involve time: 'there is simply no prior epoch in which a preceding causative agency, natural or supernatural, can operate'.[10] The quantum fluctations are not caused by anything but are spontaneous: the radioactive decay of uranium atoms.

Another complication is that the expansion of the universe occurs from nothing. How? Gravitational energy is negative and matter energy is positive, the negative energy of gravity cancels the positive energy of matter, so the total energy is zero. 'In quantum theory, the universe can tunnel through this size barrier and appear spontaneously with a size greater then the critical value.'[11] A parallel between creatio ex nihilo and that the world emerged from a vacuum could be seen. A vacuum is the lowest energy state of a system but in quantum mechanics it is a state of activity because the uncertainty principle does not allow a particle to be at rest in its lowest energy position. If it did, then we would know both its position and its momentum. Thus there is a certain necessary and irreducible 'quiveriness' involved which is called vacuum fluctuation. Nothing, in this sense, becomes something, that is, the universe, by huge quantum fluctuations. It could be understood as a sea of possibilities which was God's way of actualising the Big Bang and the beginning of the universe. But what we know about quantum fluctations occurs in our spacetime framework. How could they occur in a situation where space and time did not exist? Again, quantum physics, as we mentioned in discussing Hawking's proposal, is usually applied to the behaviour of parts of the universe. How could it be applied to the universe as a whole? The theory is speculative but has some justification, for virtual particles have been created in the laboratory.

Some scientists think that something may arise out of nothing by chance. But the probabilistic laws of the quantum need to be present so that the universe will 'fluctuate' into existence and the existence of such laws does not justify us calling it chance. The odds against this are immense.[12] Where did the quantum laws come from? Not with the universe otherwise we cannot understand them as an explanation of it. Their status has been debated from thinking of them as eternal or human inventions or ...[13] But as far as the design argument is concerned it is the rationality of the laws that point to the designer as well as the structure of the world rather than the philosophical discussion of a temporal series of cause and effect. It is the laws that permit life eventually to emerge and we express them in mathematical formulae. With regard to Hawking, his proposal does not require a background of spacetime but it is claimed that his complicated maths and imaginary numbers do not relate to real time. It is a purely mathematical idea which does not necessarily have a physical correlate for there is a difference between a mathematical construction and its physical embodiment.

The theist's belief in a creator does seem more probable than that laws and elements just happened by a cumulation process, accident and randomness. Why the laws of nature are what they are so as to bring the universe into existence is the key question for those who hope to put forward a theory of everything. But as Russell Stannard points out the difficulty is that a theory of everything (TOE) would be based on axioms that could have been otherwise and cannot be proved to be consistent from within the system. It would not be complete, for Kurt Gödel demonstrated that from within a mathematical system it was impossible to prove the truth of all true statements contained in that system. Incompleteness is true of whatever mathematical system models our universe. We belong to the physical world and we will be included as part of that model, hence we will not be able to justify the choice of axioms in the model and the physical laws to which these axioms correspond nor account for the statements made about the universe. Stannard concludes that ' … the goal of a complete theory of everything is unattainable, and the claim to have disproved the need of a creator is false'.[14]

Creatio ex nihilo means that God created order out of nothing and it is the accuracy of such order that implies design. The universe required this for its existence and continual sustaining by God. This means dependence: a mode of relationship not a primeval act. The world being contingent is not self-explanatory and the theist argues that since everything that exists depends on something else, the universe depends on a self-existent God. Creation can be seen as a giving of God, an overflow of his Spirit, a dynamic interaction, and it could involve what we mentioned earlier a self-limiting of the Being of God.

With regard to humanity the religious traditions regard the soul as an important contact with God. It can be understood as our thinking, feeling and willing, and has emerged from matter. It is the self or I which we are conscious of and persists throughout life and could survive death: the pattern in a new body. Dualism is avoided for the person is a psychosomatic unity. Being a personal agent means embodiment for how can we identify someone without it? Thus Strawson says that Christians have 'wisely insisted upon the resurrection of the body'.[15] Creation may be understood as God making matter the bearer of Spirit as signified by the sacraments of the Christian church. But the whole universe can be seen as sacramental which is one reason why the Quakers refuse the sacraments lest they restrict God's presence.[16]

What model of God then can be put forward? It will include elements in the models which we have examined such as Being in relation, Being in Becoming, Self-limitation in creation and incarnation, Transcendent and Immanent, Divine Persuader and so on. These models emerged in a social and historical context and ours must reflect the scientific and technological age otherwise we fail to communicate with it.

Identity-in-difference can help us here. If the human mind is a global attribute of the brain and not a local product of any of its parts (Chapter 6) and the rational order of the universe is akin to human reason (Chapter 1) then we have a pointer to a Mind behind it. The human mind is transcendent to the working of the brain, and analogously, we can think of a Cosmic mind or Self which transcends the universe. It is the rational principle of the universe or logos (Stoics) or sum of the divine ideas (Philo) or the designation of Christ as the eternal Word (Jo.1.1). Some see the forms of Plato in the mind of God which are then reflected in the mathematical structure of the physical world. Thus, 'the continuing conformity of physical particles to precise mathematical relationships is something that is much more likely to exist if there is an ordering cosmic mathematician who sets up the correlation in the requisite way.'[17]

Augustine was impressed by the divine wisdom expressed in numbers and he thought that they were the key to the secret of the universe. The creator was the divine Geometer and the perfection of the divine order seen in the patterns that permeate the universe. Pythagoras and Euclid developed geometry to find the uniformity and the Wisdom of Solomon refers to the divine architect as arranging all things by measure and number and weight.[18] When we consider the vast cosmos with its 2,000,000 stars and about 100,000 million galaxies operating with such precision it would appear that it is a scientific cosmic Mind that conceived it.

If this is so what is the relation between the human mind or soul and the Cosmic Mind? We cannot identify them (pantheism) or say that they are completely different (dualism). Matter is not alien to God since he became flesh and we are made in his image. Thus identity-in-difference seems to be the solution. Some have seen it even in Sankara where the gap between the empirical ego and the Self can indicate the distance between us and God. God's thoughts are higher than our thoughts so there is a difference but there must be some similarity, as we mentioned, otherwise we would not have any knowledge of them. The identity is usually expressed by saying that God is in us and the difference by his transcendence which we have interpreted as his otherness or holiness. We can think of many things being 'in' something without identification: the bird in the air, the fish in the water, and so on. Paul's mysticism appears in this connection for he continually speaks of being 'in Christ'. But it does not mean complete identification, for 'I live, yet not I, but Christ liveth in me', is followed by 'the life which I now live in the flesh, I live by the faith of the Son God', that is the 'Thou' and 'I' stand over against each other.[19]

Identity-in-difference has been seen as panentheism which is nearest to theism. Polkinghorne rejects the view but thinks it may occur in future: a full integration into the divine life. that is, an eschatological fulfilment.[20] If he means union with God, not absorption, it is in accord with our argument but Paul must have meant that it occurred in this present life and Jesus says that those who believe in him have eternal life now (Jo.3.36). We do not think that a panentheism which holds that the universe is the body of God is acceptable though it has been embraced not only by Hinduism but various Christian theologians. Are we to take it literally or metaphorically? Even if the latter is the intention it may be asked: what happens when the universe ends in the Big Crunch? What happens to God? Does he risk death too?[21] A theism which stresses not only God's transcendence but immanence in the world as sustainer seems preferable for it is saying what panentheism asserts: that God is present throughout, though not fully contained within the universe. It does not need to go on and speculate about it being part of his body. Ramanuja did but he was careful to say that it meant dependence on God which agrees with the Semitic traditions.

What we have said about the Mind of God, and we are conscious of how limited our knowledge is here, would suggest that it reflects a scientific Mind. And, if that is so, the world can be viewed as a vast experiment. It is where free creatures are given the opportunity to develop and respond to him. Peacocke thinks of the world as an experiment with God like the composer of an unfinished symphony, improvising, experimenting, expanding a theme and variations. Chance is God's radar beam sweeping the potentialities of the world and exploring the range of potential forms of matter. Polkinghorne objects arguing that it is the world which is experimenting seeking to actualise its possibilities. It is creation that discovers the potentialities with which it has been endowed, there is a divine 'letting be' not the working out of a deterministic plan. God interacts but does not overrule. But he admits that it involves risks for God in that things can go wrong and he thinks that God is vulnerable.[22] It is a precarious situation.

We would agree with Polkinghorne that God does not determine but he must keep control of the situation by influencing it and both he and Peacocke are agreed about this. But if God has created the world and controls it then it is his experiment: he uses 'chance' to explore possibilities, and necessity expressed in the natural order of the world that sets limits to it. The experiment of creating free creatures means that it may go wrong and there are signs of this: the Fall of humanity, doubts expressed about creation (Gen.6.6), the Flood (Gen.6.17) and so on. The stories do not fit a predetermined plan which traditionally has raised all kinds of questions about divine foreknowledge and free-will. God it would seem takes risks. As Hawking says he not only plays dice but throws them.

A model of God who has modified his attributes and granted freedom to us supports the view of it as an experiment. God calls for the response of faith but it is tested by trial: God permits the suffering of Job and the long list of heroes mentioned in the eleventh chapter of the Epistle to the Hebrews. These trials cannot be avoided for the world is a laboratory, a testing ground for faith, not some kind of Elysium field. The purpose of the experiment is to produce creatures who will freely respond to his love but how can they do this if they live in an ideal world such as some philosophers demand? If I lack nothing how can someone show their care and love for me? If I do not suffer pain or illness I do not need love from anyone. How can I choose the good if there is no evil alternative? There can be no acts of heroism or self-sacrifice in such a world: 'Just as in physics the notion of an infinite continuum of negative mass particles acquires meaning only when it is possible to point to the particle's opposite number, the anti-particle, so an all pervading goodness and love can only take on meaning when that perfection is broken and there is an encounter with their opposites; anti-goodness and anti-love.'[23]

The cosmic scientist wants us to respond to his love but for the experiment to have meaning a potential must exist for free creatures to reject that love. 'It is not God who creates the evil; he merely opens up the possibility of our rejecting him; it is then our act of rejection that creates the evil; evil is our responsibility.'[24] A father may give his son permission to drive the family car but he is not responsible for an accident or reckless driving which ignores the state's laws: he in no sense wills it. God wills the universe to exist and permits evil but establishes laws which if broken can result in national and personal tragedy. He calls for us to cooperate in fighting evil and will ultimately punish the wrong doer.[25]

God can only be compared to a scientist if the caricature of the scientist depicted in fiction and films as clinical, rational, impersonal, objective and devoid of personal feelings, is set aside. This was the traditional position but we have seen a move away from it to one who is personally involved in various ways. Quantum theory points to the crucial role of the observer and in relativity the measurements of mass, velocity and length of an object depend on the frame of reference of the observer. The scientist has to exercise personal judgement on the evidence that he accumulates. It is true that personal involvement is more clearly seen in religion than science but scientific data is theory laden, there is selection and reporting, interpretation of data and creative imagination. The scientist cannot escape values. And it has been shown in the social sciences that knowledge is often gained by participation. Thus David Hardgreaves, when studying the comprehensive school, became a teacher in it and sought a sympathetic rapport with the pupils and Alexander Whyte became a member of a street gang in order to understand their beliefs and motivation.

This makes the social scientist a more appropriate analogy, for he interacts with the human in a way that the physical scientists cannot do but biologists have established intimate contact with chimpanzees and the human qualities of animals are more respected today. It is through interaction that knowledge is obtained and the same applies in connection with God. He is known by what he does. But could we pray to some cosmic scientist? Does it not conflict with what we have said about him as father? The father image reflects God as the personal object of prayer but the psalmist who communes with him holds it in tandem with his view of the cosmic dimensions of deity as he surveys the world which God has created. It leads him to bow in prayer before such a God. One image does not exclude the other. Krishna in Hinduism is father, friend and husband, but it does not exclude the vision of power and majesty revealed to Arjuna who worships such a revelation.

Subjective feelings are evident in the scientist when we observe the elation which occurs in the laboratory after a successful experiment. It parallels God's joy when his creation responds to him. The scientist is compassionate with the subjects involved for her aim is to benefit mankind and cause as little suffering as possible. But it is in the suffering of the scientist that we see clearly his involvement. Derision and scorn greeted Copernicus when he asserted that the earth was moving and he was put on trial and Galileo was threatened with torture. Newton neglected his health in writing the *Mathematical Principles of Natural Philosophy* and the results of 20 years work in optics were lost in a fire. In order to understand the nature of light he stared at the sun until he almost went blind and stuck blunt needles behind his eyeballs. His sight only properly returned after a period in a dark room. He wrote extensively on alchemy and tasted a variety of metals and toxic substances which may have unbalanced his mind.[26]

Lavoisier, the great chemist, was accused of stealing other men's ideas, thrown into prison in 1793, and sentenced to death. When the Court was told that, in the opinion of most of the scientists of Europe, citizen Lavoisier occupied a distinguished place among those who had brought honour to France, the President replied: 'the Republic has no need of men of science'.[27] In medical science, Lister had to fight for antiseptics in surgical operations and Simpson battled against opposition in the use of chloroform. In our century 200,000 people died as a result of the atomic bomb being dropped on Hiroshima and Nagasaki. J. Robert Oppenheimer in 1949 struggled to get the military to agree to a demonstration of its power with Japanese observers present. This was opposed on the grounds that it was moral to do what they had done to the

Americans: the bombing of Pearl Harbour without warning. Oppenheimer refused to work on the hydrogen bomb and in 1954 his security clearance was revoked and he was forced from public office. He died of cancer in 1967.

Scientists risk death. Marie and Pierre Curie worked for years to discover radium. They experimented with pitchblende trying to isolate what they thought might be radium and one night they watched with awe the strange glowing of radium in the darkness of their old shed. They did not know the risks they were running and Marie Curie's exposure to radium resulted in twisted and deformed hands and leukaemia and she died prematurely because of such radium absorption. One of the achievements of the Curies was radiation to treat cancer, a form of which cost Marie Curie her life.

The Cosmic Scientist is also engaged in warfare against disease and death and the agent and incarnational models show the suffering involved. T. S. Eliot uses the image of God as 'the wounded surgeon' and others speak of the vulnerable God.[28] Genesis refers to God having to rest on the seventh day. Does this mean that God like a surgeon in a very difficult operation was drained by the effort poured out and into his creation? But it is the incarnation of God in Christ which demonstrates in particular the entry of God into the experiment, the participation in human life, the suffering involved in the atonement, and the extent of evil that has to be defeated. The cosmic dimension of such a conflict emerges in the writings of St Paul and the victory of the Cosmic Scientist over it (Col.2.15). The religious picture complements the scientific since it not only gives us his majesty and mystery but his wisdom, justice, love, holiness, truth and goodness.

Stephen Weinberg calls for a God who is involved in the world and has established standards of good and evil. Values have been seen as the goal of the evolutionary process and we mentioned in Chapter 6 that philosophy understood them as justice, generosity, truthfulness, kindness, love, faithfulness and compassion.[29] The religious traditions agree. Islam follows Judaism in stressing the love of the neighbour and the duty of helping the widow, the orphan, the sick and the poor. Christianity asserts that love for God and for the neighbour are the basic requirements. The Buddhist emphasises compassion and the Sikhs oppose stealing, murder and adultery and only resort to force in the face of evil. The sacredness of life in Hinduism is well illustrated by Gandhi's non-violence principle based on both the Gita and the Sermon on the Mount. Beauty, truth and goodness oppose unjust competition and the view that we are just survival machines. But the struggle to achieve these values is bought at the price of suffering as the scientist is well aware when the truth he or she has discovered is opposed. But unless the right values operate both in science and religion neither will benefit society. God is not indifferent to the suffering involved in achieving the standards which he demands for if the incarnation model is correct he has entered into the experiment and shown how they can be achieved.

In conclusion we note that we have put forward suggestions as to what God is like which are tentative and probable. We do not think this is odd because even the humble electron is difficult to define. We know what it does in resisting motion, attracting and repelling other particles but not what it is. We base our knowledge of what God is like on his interaction with us but when we try to understand the experience and who has given it the difficulties arise. We form our theories and models but recognise their limitations. Electrons and masses are known by their effects on other entities and their relationships. God too is known by the change in our lives as shown by our concern for others. While the model we have put forward is only probable it is comprehensive drawing on a wide range of data, it is rational, agrees with the theistic strand in the religions, and fulfils Weinberg's demand for a God involved in our lives.

The experiment will be a success not only in the defeat of evil as shown by the resurrection of Christ but in the millions of every race and creed who have responded to the love of God through the ages and have tried not only to understand his ways and interpret the world he has created but also to cooperate with him in changing it.

Notes

CHAPTER 1

1. J. G. Brunton, *The Story of Western Science*. CUP, Cambridge, 1966, p.8ff
2. Z. Sardar and Z. Malik, *Muhammad*. Icon Books Ltd, Cambridge, 1994, p.98ff
3. Alan Richardson, *The Bible in the Age of Science*. SCM, London, 1961, p.14
4. Ibid p.16
5. Another reason was the Greek lack of interest in technology. Manual work was done by slaves who were less expensive than building waterwheels. Concerning education the platonic curriculum begins with physical training, literature, and then mathematics which tests mental ability. The education divides society into the workers, warriors and rulers. Only those who can cope with higher mathematics and philosophy can become rulers. Plato believed that the world has a underlying mathematical structure waiting to be discovered. It is in this way that scientists express the laws of the universe.
6. W. Wordsworth, *Lines Composed a Few Miles above Tintern Abbey, Oxford Anthology of English Poetry*. OUP, Oxford 1940
7. J. G. Brunton, op. cit. p.13
8. Ian Barbour, *Issues in Science and Religion*. SCM, London, 1966, p.30
9. Ibid p.28
10. Ibid p.21f. Aquinas was worried about understanding the scripture too literally. How were the anthropomorphic pictures of God in the Bible to be reconciled with his transcendent majesty, holiness and glory? He used analogy to postulate both similarity and difference between God and us. If a non-literal interpretation had been heeded the ensuing disagreement between Galileo and the Church might not have occurred. But there was also a clash of personalities as the pope thought Galileo was discrediting his views.
11. A. Richardson, op. cit. p.17. Aquinas had adapted Aristotle and Ptolemy so well to Christianity that when their cosmology was attacked it seemed an assault on the faith. Aristotle had taught that the motionless earth was surrounded by fifty revolving crystalline spheres, carrying the sun, the moon, the planets and the stars. The outside sphere was moved by God and beyond it there was nothing. But this movement of God had now disappeared with the new view of the universe. Ptolemy had explained more satisfactorily then Aristotle the variation in distance from the earth of the planets but he also believed in a motionless earth. The opposition to Copernicus and Galileo was led by Aristotelian philosophers and theologians who knew that it was a challenge to their authority and that of the establishment.

12. I. Barbour, op. cit. p.25
13. Newton's laws are well known and do not need stating: they appealed to common sense. But since Einstein was to deal with gravity we mention Newton's view. It causes bodies to fall to earth, causes the tides, holds the moon in orbit round the earth, preserves satellites in orbit around their planets and maintains the planets in orbit around the sun. It steers the comets so it applies outside the solar system. All bodies are endowed with a principle of mutual gravitation, every two bodies gravitate towards each other in proportion to their masses and in inverse proportion to their distances. Newton's method was to reduce nature to general rules or laws and establish them by observations and experiments and deduce the causes and effects of things. In 1846 Neptune was discovered and confirmed Newton's laws. William Rankin, *Newton*. Icon, Cambridge, 1993, p.148f
14. A. Tilby, *Science and the Soul*. SPCK, London, 1992, p.43
15. Ian Barbour, op. cit. p.38
16. Ian Barbour, op. cit. pp.43, 15. *Religion in an Age of Science*. SCM, London, 1990, p.220. Roger Scruton, *Modern Philosophy*. Sinclair Stevenson, London, 1994, p.46
17. John Houghton, *The Search for God, Can Science Help?* Lion Publishing, Oxford, 1995
18. David Goodman, 'Muslims and Jews in Europe', A205 course, *Culture and Belief in Europe, 1450–1600*. Open University Press, Milton Keynes, 1992, p.20ff

CHAPTER 2

1. The autobiography of Charles Darwin (1887). Collins, London, 1958, p.28. This edition restores omissions. His father said to him: 'You care for nothing but shooting, dogs, and rat-catching, and you will be a disgrace to yourself and all your family'.
2. Ibid p.85
3. A. Desmond and J. Moore, *Darwin*. Penguin, London, 1992, p.176
4. Ibid p.191
5. C. Darwin, op. cit. p.87
6. Ibid p.92
7. Ibid p.121
8. *The Descent of Man*. John Murray, London, 1871. vol. i. p.65ff
9. A. Desmond and J. Moore, op. cit. p.148
10. J. G. Brunton, *The Story of Western Science*. CUP, Cambridge, 1966, p.101
11. OU Course Team, *The Nature of Evolution* unit 1 S364. Open University, Milton Keynes, 1981, p.10
12. B. S. Beckett, *Biology: A Modern Introduction*. 2nd edn. Oxford University Press, 1982, p.238. Darwin never really solved the problem of inheritance which was only clarified through the work of Mendel, Morgan, Weissman, and the advance of study of genetics. In later

editions of the *Origin of Species* he accepted the inheritance of acquired characteristics postulated by Jean Baptiste de Lamarck (1744–1829). An example is the giraffe. How did it get that long neck? Lamarck held that its short-necked ancestors were striving to reach the leaves of trees and must have stretched their necks. The acquired characteristic was inherited by their young who continued the stretching and passed it on. The hypothesis is doubtful though recent investigations appear to see some truth in it.

13. OU units, S364 op. cit. p.8
14. A. Desmond and J. Moore, op. cit. p.244
15. Darwin, op. cit. p.200, vol.2
16. Ibid vol. 1, p.106. C. Ralling (ed.) *The Voyage of Charles Darwin.* Ariel Books, BBC, London, 1978, p.84
17. M. Denton, *Evolution: A theory in crisis.* Burnett, London, 1985, p.56
18. C. Darwin, op. cit. vol. 1, p.201
19. C. Darwin, op. cit. p.162
20. *Religion in Victorian Britain.* University Press, Manchaster, 1988, 11 p.179
21. J. M. Golby, *Culture and Society in Britain 1850–1890.* OUP, Oxford, 1986, p.58
22. Richard Dawkins, *The Blind Watchmaker.* Longman, London, 1986, p.91
23. M. Denton, op. cit. p.61
24. J. D. Barrow and F. J. Tipler, *The Anthropic Cosmological Principle.* Clarendon Press, Oxford, 1986, p.87. Huxley insisted that the only way to truth was by scientific investigation. But he thought that atheism was untenable and coined the word agnostic. He believed that the Bible should be taught in the schools because of its literary merit and as a ground of ethics. He was seriously perplexed, 'to know by what practical measure the religious feeling, which is the essential basis of conduct, was to be kept up, in the present utterly chaotic state of opinion in these matters, without its (Bible) use.' EB vol. 2 William Benton, London, 1963, p.948
25. Darwin, op. cit. vol. 1, p.166
26. Quoted by O. Hardy, *Darwin and the Spirit of Man.* Collins, London, 1984, p.76
27. S. Gould, *The Panda's Thumb.* Penguin, London, 1980, p.50ff
28. A. Desmond and J. Moore, op. cit. p.570. Wallace had a different opinion of the natives than Darwin. He had been in the Amazon and the Far East and discerned at times a higher morality than the colonists who often tried to exterminate them. As far back as the 16th century travellers arrived at the same conclusion when they saw how so called Christian nations behaved in their conquest of the Indies. Wallace insisted that the cut-throat capitalism of Victorian England demonstrated that natural selection did not have the power to improve morality or enhance civilisation. His reply to Darwin's criticism was that he had come to the conclusion after reflecting on the existence of forces and influences not yet recognised by science. Both Wallace and Lyell saw the presence of God in evolution.
29. I. Barbour, *Issues in Science and Religion.* SCM, London, 1966, p.96
30. A. Desmond and J. Moore, op. cit. p.387

31. A. Desmond and J. Moore, op. cit. p.636
32. J. R. Moore, 'Freethought, Secularism, Agnosticism; The case of Charles Darwin', p.313 in *Religion in Victorian Britain*. University Press Manchester, 1988, vol.1
33. Michael White and John Gribbin, *Darwin: A Life in Science*. Simon & Schuster, London, 1995; J. Greer, op. cit. p.23
34. J. R. Moore, op. cit.
35. I. Barbour, op. cit. p.90f
36. C. S. Lewis, *The Problem of Pain*, argues that evolution could have taken a wrong turn.
37. I. Barbour, op. cit. p.91. Gray claimed that a Darwinian teleology had an advantage over the more traditional version of the argument from design for evolution showed that apparent wastage in nature is an indispensable part of the process. Without it mankind would never have come into existence at all: good requires evil to combat and suffering can be explained in this way. But Darwin was pessimistic because of the suffering of the innocent. The answer that the world might be a moral training ground did not impress him. He was not prepared to accept the doctrine of the Fall of mankind and its consequences for both man and nature (Rom. 8:22).
38. I. Barbour, op. cit. p.93. With regard to values in the Victorian Age, G. Best argues that there was a consensus of values up to 1870 with deference, respectability and independence. There was also self-help, duty, responsibility, individualism and the sacredness of work. But he contended that there was a break up of the consensus from 1870 which has been debated. G. Best, *Mid-Victorian Britain (1852–1875)*
39. A. Richardson, *The Bible in the Age of Science*. SCM, London, 1961, p.91
40. Ibid p.71. The Protestant authority was scripture and it was the Bible which was subjected to the assault of Darwinism and historical criticism. The authority of Catholicism was the Church and it suppressed such criticism. Consequently, as the 19th century proceeded both in England and in Europe the Protestant authority shifted to religious consciousness and ethical values. Religion tended to become a matter of feeling and was in danger of losing objectivity. The masses thought that the established church was supporting the status quo with its unequal and oppressive class system. But the Catholic Church succeeded in the ghettos with their large influx of Irish immigrants and deserved the name of being the church of the poor. G. Parsons, 'Victorian Roman Catholicism; Emancipation, Expansion and Achievement' in *Religion in Victorian Britain*. University Press Manchester, vol.1, p.178f. Also vol. 2, 1988, p.77ff. In October 1996 the pope acknowledged the theory of evolution though he still regards the soul as of immediate divine creation.

CHAPTER 3

1. F. Crick, *Life Itself*. MacDonald & Co., London, 1982, p.18f
2. Jason Burke, *The Sunday Times*, 8th Oct. 1995

3. A. Scott, *Vital Principles.* Blackwell, Oxford, 1988, p.185
4. P. Davies, *The Cosmic Blueprint.* Heinemann, London, 1987, p.139
5. Quoted by John Greer, *Evolution and God.* CEM, London, 1979
6. R. Dawkins, *The Blind Watchmaker.* Longman, Essex, 1986, p.158. Alan Hayward referring to Fred Hoyle's *Universe Past and Present Reflections, Life and Work,* Oct. 1995, p.17ff. In recent times there has been speculation about life on Mars due to the discovery of a 13,000 year old meteorite in Antarctica. The rock revealed tiny worm-like objects resembling microfossils of bacteria and tiny deposits of carbon compounds, an indication of life. Scientists are cautious about conclusions but Fred Hoyle said that it might confirm his theory. The theist will argue that it only shows that God has created a much greater universe and he is bound to be concerned about it.
7. R. Hubbard, *Exploding the Gene Myth.* Beacon Press, Boston MA, front page
8. Ibid p.3f, p.128. Ruth Hubbard points to attempts at correlating genes and behaviour: the link between XYY chromosomes amd extreme aggression. Research was carried out in a Scottish mental hospital where it was estimated that the inmates had about 20 times the proportion of XYY than the general population. But it showed that most had not committed crimes of violence. Studies in the population demonstrated that most XYY males are not excessively aggressive. Ibid p.104ff, p.145ff. The most extensive study of the criminal gene was begun at Camberwell, London in 1961. More recently, Dr Han Brunner, a Dutch scientist, claimed that he has isolated the criminal gene. But a gene is just a chemical that shows its effect in particular combinations and environments. *Radio Times,* 1–7 June 1996, p.29, BBC, London
9. William Bains, *Genetic Engineering for almost Everybody.* Penguin, London, 1987, p.18
10. R. Hubbard, op. cit. p.45
11. R. Dawkins, *The Selfish Gene.* OUP, Oxford, 1989, p.28
12. John Bowker, *Is God a Virus?* SPCK, London, 1995, p.7. E. O. Wilson, *On Human Nature,* 1978
13. Ibid p.29ff
14. Ibid p.41f, p.72
15. Ibid p.74
16. R. Dawkins, op. cit. p.198
17. Quoted by J. Polkinghorne, *One World.* SPCK, London, 1986, p.28
18. R. Dawkins, *The Blind Watchmaker.* Longman, Essex, 1986, p.21
19. J. Polkinghorne, *Science and Providence.* SPCK, London, 1989, p.38
20. J. Polkinghorne, *Science and Creation.* SPCK, London, 1989, p.48. Polkinghorne writes: 'The raw material of novelty thus provided by chance is then explored by the intervention of lawful necessity to sift and preserve those configurations which manifest their fruitfulness by their survival and replication in a regularly behaving environment … Thus the potentiality of the universe is brought to actuality …'. Random events can prove to be the originators of pattern.
21. J. Polkinghorne, *Scientists as Theologians.* SPCK, London, 1996, p.45
22. R. Dawkins, op. cit. p.91

23. Ibid p.141. Dawkins generated programs on his computer to give him a random phrase of 28 letters and then set up a goal phrase taken from Hamlet: 'Methinks it is like a weasel'. After many sequences of the letters and many mistakes, a closeness to the phrase was reached and he then compares the mistakes with the mutations in the evolutionary process. The operation required forty three generations to reach the phrase, but Dawkins was the designer with a goal in mind: the phrase, and the computer was programmed to choose a winner on the basis of resemblance to that goal design. In any case how can the computer state which is static be compared with the development of life and a changing world?

24. Quoted by R. Sheldrake, *The Presence of the Past.* Collins, London, 1988, p.84

25. Ibid p.87.

26. Paul Davies, op. cit. p.102ff

27. Niles Eldredge, *Reinventing Darwin: The great evolutionary debate.* Weidenfeld & Nicolson, London, 1995, p.63ff, 105, 178, 226f. J. Bowker, op. cit. p.30f

28. Eldredge, op. cit. p.184. Dawkins, John Maynard Smith and George Williams are regarded as ultra-Darwinians (theoretical geneticists) while Gould, Dick Lewontin, Steven Rose and Eldredge are called 'naturalists'.

29. S. Gould, *The Panda's Thumb.* Penguin, London, 1980, p.77

30. Ibid p.159. Viruses, parasitic strands of DNA or RNA (ribonucleic acid), can drift free of living cells, acquire hard protein shells, and behave like pollen-bearing bees, flitting from cell to cell and carry genes. They invade a cell, insert their own DNA and trick the cell's reproductive machinery into manufacturing new viruses. These RNA retroviruses, may become part of that species' heredity and can cause deadly diseases: it has been suggested that one may have jumped from African green monkeys to infect people with Aids. It has been called the 'jumping gene'.

31. R. Dawkins, *The Selfish Gene.* p.201

32. J. Bowker, op. cit. p.33

33. Andrew Scott, *Vital Principles.* Basil Blackwell, Oxford, 1988, p.95

34. John Cornwell, 'Chance is fine thing', *New Scientist,* vol. 150, No. 2027, 27th April 1996, p.47. Cornwell thinks that Dawkins' work would gain a lot by acknowledging the diversity of views.

35. M. Fuller, *Atoms and Icons.* Mowbray, London, 1995, p.33. Ibid p.34

36. Karl Popper, World Philosophy Congress in Brighton 1988, *Guardian* 29 Aug. 1988. Compare Bronowski, *The Common Sense of Science.* Heinemann, London, 1982 edn, p.72f

CHAPTER 4

1. J. Schwartz and M. McGuinness, *Einstein.* Icon Books, Cambridge, 1992, p.23ff, p.137ff

2. Danah Zohar, *Through the Time Barrier.* Heinemann, London, 1982, p.115

3. J. Schwartz and M. McGuinness, op. cit. p.48
4. Ibid pp.59–62
5. R. W. Clarke, *Life and Times of Einstein*. 1973, p.87
6. J. Schwartz, op. cit. p.83
7. Ibid p.98
8. J. Schwartz, op. cit. p.4
9. J. Schwartz, op. cit. p.109
10. D. Zohar, op. cit. p.118
11. R. Stannard, *Doing Away with God*. HarperCollins, London, 1992, p.31
12. Ibid
13. R. W. Clarke, op. cit. In Euclidean geometry, the angles in a triangle equal two right angles. But a triangle formed by the Equator and two lines of longitude, that is, running from the Equator to the North Pole through Greenwich and New Orleans, encloses with the Equator, not two but three right-angles! Einstein said that when the blind beetle crawls over the surface of a globe, he does not know that the track he has covered is curved. It is the curvature of light in a gravitational field that requires Riemannian geometry. With it parallel lines do not exist and the angles of a triangle do not add up to 180 degrees. Perpendiculars to the same line converge and the shortest lines joining any two points are not straight. The shortest distance between any two points on a curved surface is worked out by a formula different from the length of a line on a plane surface. Einstein used this geometry to understand the movements of the stars and the universe itself. He spoke of 'curved space'. This has nothing to do with the shape of space, bent or not, but relates to the way 'distance' is defined. It is not the space that is curved but the geometry of the space. But the basic point is that light did not go straight and the universe can only be viewed from the earth through the distorting spectacles of gravity. The planets followed the same elliptical paths round the sun with only insignificant change but Mercury at the point on its elliptical path which was nearest the sun advanced by a specific amount each year.
14. R. Stannard, op. cit. p.28
15. J. Bronowski, *The Common Sense of Science*. Heinemann, London, 1951, p.67
16. R. Clarke, op. cit. p.225
17. A. Tilby, *Science and the Soul*. SPCK, London, 1992, p.85
18. J. McEvoy and O. Zarate, *Stephen Hawking*. Icon Books, Cambridge, 1995, p.70ff
19. Ian Barbour, *Religion in an Age of Science*. SCM, London, 1990, p.110
20. Ibid p.111
21. Ibid p.111
22. Peter Millar, *The Sunday Times*, Oct 8, 1995
23. D. Zohar, op. cit. p.119f
24. Stephen Hawking, *Black Holes and Baby Universes*. Bantam Books, London, 1993, p.36
25. R. Clarke, op. cit. p.502

26. J. Polkinghorne, *Science and Providence*. SPCK, London, 1989, p.82
27. A. Tilby, op. cit. p.70
28. S. E. Frost, *Basic Teachings of the Great Philosophers*. Dolphin Books, New York, 1962, p.116f
29. Ian Barbour, *Issues in Science and Religion*. SCM, London, 1966, p.434ff
30. John Houghton, *The Search for God. Can Science Help?* Lion, Oxford, 1995, p.132
31. Stephen Hawking, op. cit. p.69
32. Peter Lewis referring in *Daily Mail*, May 25, 1996 to the biography of Einstein by Denis Brian, *Einstein*. John Wiley, London, 1996

CHAPTER 5

1. S. Hawking, *Black Holes and Baby Universes*. Bantam Press, London, 1993, p.102
2. J. Bronowski, *The Common Sense of Science*. Heinemann, London, 1951, p.74
3. I. Barbour, *Issues in Science and Religion*. SCM, London, 1966, p.302
4. J. Polkinghorne, *One World*. SPCK, London, 1986, p.45
5. Ian Barbour, op. cit. p.281
6. R. Stannard, *Science and the Renewal of Belief*. SCM, London, 1982, p.112
7. Ibid p.194f. Before a pair of quantum particles is measured they are in a fuzzy state but if one is measured the state of the other is defined even if far apart. There is an intimate bond yet no signal has passed between them. Bohr's view was confirmed in 1982 by Alain Aspect in the Institute of Optics in Orsay near Paris. *New Scientist*, 28 Sept. 1996, p.27. The effect can now be put to work in machines. An electron in an atom has no definite position only possible locations each described by a different quantum state so there is only probability that the electron is in one of these states.
8. J. Bronowski, op. cit. p.95
9. Thanu Padmanabhan, 'Quantum cosmology and the Creation', in *Frontiers of Science* ed. A. Scott. Blackwell, Oxford, 1990, p.153ff. Suppose we roll a dice 6000 times and count the number of times the '3' turns up. We might expect 1000 times but every time we try we may get results such as 983 or 1004, or 976 or ... The average value will be 1000 but there will be fluctuations around the average value. If we measure the position or speed of an electron we will get a run of numbers which have an average value but they will also exhibit fluctuations around the value. Hence we can only compute the probability that the system will do various things.
10. F. Tipler, *The Physics of Immortality*. Macmillan, London, 1995, p.211
11. I. Barbour, *Religion in an Age of Science*. SCM, London, 1990, p.111
12. M. Fuller, *Atoms and Icons*. Mowbray, London, 1995, pp.24,18
13. S. Hawking, *Black Holes and Baby Universes*. pp.73,14
14. S. Hawking, *A Brief History of Time*. Bantam Books, edn 1995, p.47

15. Ibid p.144
16. R. Stannard, *Doing Away with God*. HarperCollins, London, 1993, p.103
17. J. P. McEvoy and O. Zarate, *Stephen Hawking*. Icon Books, Cambridge, 1995, p.49
18. S. Hawking, *Black Holes and Baby Universes*. p.45
19. J. P. McEvoy, op. cit. p.165
20. P. Davies, op. cit. pp.203f, 232. J. D. Barrow and F. J. Tipler, *The Anthropic Cosmological Principle*. Clarendon Press, Oxford, 1986, p.22
21. Paul Davies, *God and the New Physics*. Dent, London, 1983, p.55f
22. S. Hawking, op. cit. pp.102–108
23. J. P. McEvoy, op. cit. p.117f
24. Ibid p.137
25. S. Hawking, *Black Holes and Baby Universes*. p.84
26. Ibid p.76
27. Ibid p.84
28. J. P. McEvoy, op. cit. p.151f
29. Ibid p.166
30. A. Tilby, *Science and the Soul*. SPCK, London, 1992, p.121
31. S. Hawking, *Black Holes and Baby Universes*, op. cit. p.158
32. Ibid
33. P. Davies, *The Mind of God*, Simon & Schuster, London, 1992, pp.61, 158
34. Alvin Toffler (Ed.), *Introduction to Order out of Chaos* by Ilya Prigogine and Isabelle Stengers. Fontana, London, 1984, p.xvi. According to the second law of thermodynamics there is in the universe a loss of energy capable of doing work. The universe is running out of energy and moving from order to disorder: an increase in entropy. But with regard to systems in the world, order can emerge out of the disorder for while some do run down others evolve and become more organised.
35. B. Russell, *Mysticism and Logic*. Allen & Unwin, 1971, p.9f
36. J. Polkinghorne, op. cit. p.94

CHAPTER 6

1. David Bohm, *Wholeness and the Implicate Order*. Routledge & Kegan Paul, London, 1980, p.51
2. R. Penrose, *Shadows of the Mind*. Vintage, London, 1994, pp.19, 45, 52ff
3. S. Rose, *The Making of Memory*. Bantam, London, 1992, pp.3, 87, 317f
4. J. Weizenbaum, *Computer Power and Human Reason*. Penguin, Middlesex, 1976, p.208f. A good discussion of computers, artificial intelligence and the human soul, is in *God and the Mind Machine*, SPCK, London, 1996, p.135f, by John Puddefoot. He argues that mind grows distinguishing it from the artificial intelligence of the most advanced machines.
5. Ibid
6. G. C. Davenport, *Essential Psychology*. HarperCollins, London, 1992, p.70

7.   O. Hanfling, *Body and Mind*, A313 units 1–2, Open University, Milton Keynes, 1980 pp.45, 60
8.   G. C. Davenport, op. cit. p.126
9.   Ibid p.92f
10. Roger Penrose, *The Emperor's New Mind*. OUP, Oxford, 1989, p.426. He cites the case of a chimpanzee who was closed in a room with a box and a banana suspended from the ceiling, just out of his reach. After a number of vain attempts he appeared to think about the problem. His eyes moved from the banana to the empty space beneath it on the ground, from this to the box, then back to the space, and from there to the banana. Suddenly he gave a cry of joy, and somersaulted over to the box in high spirits. He pushed the box below the banana and reaching up got the prize.
11. R. D. Gross, *Psychology*. Hodder & Stoughton, London, 1991, p.182
12. R. D. Gross, op. cit. p.32
13. R. D. Gross, op. cit. p.195. Behaviourist psychologists dismiss thinking as simply talking to ourselves. But in an experiment (1947) the claim was questioned. One of the experimenters, Smith, allowed all his bodily movements to be paralysed by a drug and was kept alive on a respirator. According to the behaviourists he should have been unable to think since he could not move. But Smith clearly recalled his thoughts, feeling and experiences during the experiment. It was also shown that some patients wake during anaesthetic and are aware of what is going on but are unable to communicate.
14. R. Popkin and A. Stroll, *Philosophy*. Doubleday, New York, 1956, p.185
15. R. Popkin and A. Stroll, op. cit. p.76
16. Alun Anderson, 'Zombies, Dolphins, and Blindsight', *New Scientist*, 4th May 1996, p.20f
17. F. Crick, *What Mad Pursuit*. Weidenfeld & Nicolson, London, 1989, p.115
18. Ibid pp.149–63. R. L. Gregory, *Mind in Science*. Weidenfeld & Nicolson, London, 1981, p.557
19. Roger Scruton, *Modern Philosophy*. Sinclair Stevenson, London, 1994, p.550. J. Searle agrees with Penrose, and refers to the analogy of the computer. A computational role can be assigned to a system but only if it used as such. The question arises: Who is using this brain in such a way? He supposes that I am alone in a room with a set of instructions which tell which cards to pass to those outside in response to their cards. The cards carry Chinese characters and the instructions demand that I must pass out answers in that language though I do not understand it. A computer could be programmed to do it and give all the correct responses to a given input without understanding but a person must understand. Perhaps Searle needs to take more account of the advances in computers but he is right about the mind not simply responding to input from the outer world but being perceptive and creative.
20. M. Fuller quoting R. Stannard, op. cit. p.43
21. P. Singer, *Practical Ethics*. Supplementary Material 2 OU course A310, p.21

22. A. Thatcher, *Truly a Person, Truly God*. SPCK, London, 1990, p.109. He has an enlightening discussion of person. See in particular chapters 7, 8 and 9 based on P. F. Strawson, *Individuals*, Methuen, London, 1990, p.101f

23. R. D. Gross, op. cit. pp.219, 224ff, 680

24. *Church of Scotland Report on Human Genetics*, Edinburgh, 1995, pp.12, 29, 55. The Clothier and Nuffield Reports state that limits must be placed on genetic screening, testing and therapy. Pressure should not be put on prospective parents to undergo tests or improper use made of genetic information by insurance companies and employers. Genetic testing before marriage could affect the biological normality of reproduction by creating excessive anxiety. Counselling, information and proper consent are required and commitment to the vulnerable and disabled must not be eroded.

25. G. C. Davenport, op. cit. pp.29, 271

26. G. C. Davenport, op. cit. p.34

27. R. D. Gross, op. cit. p.549. An example was of a woman dying from cancer. One drug might save her, a form of radium stocked by a druggist, but he was charging ten times what it cost to make. The sick woman's husband, Heinz, went to everyone he knew to borrow the money, but only had half the price. He begged the druggist to sell it cheaper or let him pay later but he refused and in desperation he broke into the shop and stole it. The druggist was opposing the values of generosity and justice but it was a hard case and we cannot generalise from it to set aside rules which operate in normal situations. Both Piaget and Kohlberg understood cognitive and moral development as having biological roots which meant a predisposition just as with language.

28. J. Lewis, op. cit. p.142

29. D. Scully, *God and Reason*. CEM, London, 1989, p.65. He says the acceptance of the natural principle of the survival of the fittest has meant competition, greed, self-preservation, leading to domination, power and wealth, the exploitation of the third world, poverty and debt, and the depletion of the earth's resources.

30. T. Beauchamp and J. Childress, *Principles of Biomedical Ethics*, 4th edn. OUP, Oxford, 1994, p.503

31. S. Hawking, *Black Holes and Baby Universes*. Bantam, 1994, p.12

32. J. Polkinghorne, *Reason and Reality*. SPCK, London, p.76

33. R. Stannard, op. cit. p.29

34. Other distinctions between humanity and the sub-human are clear. The dexterity of the human hands and the ability to make sophisticated tools including the computer. Our technology, ability to live in any kind of environment, domestication of animals, achievements in medicine and surgery, astronomy and mathematics, art, music and sculpture, and reaching out beyond the planet into space. Man is a unique animal for while there is continuity between the human and the sub-human there is discontinuity to a vast degree.

35. J. Weizenbaum, op. cit. p.127

CHAPTER 7

1. Ninian Smart, 'Hindu Patterns of Liberation' unit 3, OU Press, Milton Keynes, 1987, p.28
2. N. Smart, *Reasons and Faiths*, Routledge and Kegan Paul, London, 1958, p.907. Ibid p.16f
3. K. Ward, *Images of Eternity*. Darton, Longman and Todd, London, 1987, p.27
4. N. Smart, op. cit. p.54 footnote
5. B. Kumarappa, *The Hindu Concept of the Deity*. Inter-India Publications, Delhi, 1979, p.185
6. Ibid p.208
7. N. Smart, *The Religious Quest*, units 3 and 4, OU Press, Milton Keynes, p.30
8. Ibid p.25
9. Robert McDermott, 'Modern Hinduism: Gandhi and Aurobindo', unit 8, *Man's Religious Quest* AD208, OU Press, Milton Keynes, 1978, p.11
10. J. Hick, *An Interpretation of Religion; Human Responses to the Transcendent*, Macmillan, 1989
11. J. Masson, 'The noble Path of Buddhism' A228, *The Religious Quest*, OU Press, Milton Keynes, Units 9–11, p.22
12. Walpola Rahula, *What the Buddha Taught*. Wisdom Books, London, 1990, p.22. R. C. Zaehner, *Encyclopedia of Living Faiths*. Hutchinson, London, 1979, p.280
13. J. Masson, op. cit. p.32
14. W. Foy, *Reader Religious Quest*. Croom Helm, London, 1978, p.177
15. F. H. Cook, 'Zen and the Problem of Language', Ibid p.66f
16. J. H. Whittaker, 'Zen, Realism vs. Non-Realism, and Soteriology', Ibid p.78
17. H. Kung, *Christianity and World Religions*. SCM, London, p.392
18. M. Abe, 'God and Absolute Nothingness', in *God, Truth and Reality*, ed. A. Sharma, Macmillan, London, 1993, p.33ff. The Madhyamika school taught that ultimate reality is beyond description but could be reached through meditation. Objection to their negativism was brought by other schools such as the Yogacara which stressed mind in connection with the Absolute. They were idealists and thought that the external world is a product of consciousness. Nirvana is a state in which the dualism inherent in consciousness is transcended and we experience a state of pure awareness without an object. This state is the same as Emptiness. N. Smart, *The Religious Experience of Mankind*. p.138f
19. S. Radhakrishnan (ed.), *Selections from the Sacred Writings of the Sikhs*. Allen and Unwin, London, 1960, p.82
20. W. Foy, *Reader Religious Quest*. Croom Helm, London, 1978. But despite Nanak's protest about pilgrimages his followers inaugurated festival days and so on. The overall picture of patient waiting on God in meditation does not fit the usual imagery of martial Sikhism but

they had to arm against the assaults of the Mughals and form the Khalsa brotherhood (1699). It has the code of the 'Five Ks': uncut hair, comb, dagger, bangle and wearing a variety of breeches which must not reach below the knee. Prohibitions are abstinence from tobacco, from meat slaughtered in the Muslim fashion (halal), and sexual intercourse with Muslim women.

21. Terry Thomas, 'Sikhism: the Voice of the Guru', Milton Keynes, 1981, units 12–13, *Man's Religious Quest*, AD208, p.28
22. Terry Thomas, Units, p.55f

CHAPTER 8

1. J. A. Dearman, 'Insights', *Journal of the Faculty of Austin Seminary*, Texas, Spring 1995, p.23ff
2. David Goldstein, 'The Religion of the Jews'. *The Religious Quest*, A228 units 10–11, OU Press, Milton Keynes, 1987, p.14ff. The orthodox Jews reject modern historical scholarship which maintains that the Pentateuch (first five books of the Bible) is by a number of writers but the Liberal and Reform Jews argue that God has continually revealed himself throughout their history and not given one complete revelation at Sinai. The covenants (berit) or agreements which Yahweh made with Israel are an original way of conceiving of the divine-human relationship. They agreed to commit themselves to the covenants but failed to obey hence the prophets predicted an internal covenant: 'I will put my law in their minds and write it on their hearts (Jer. 31–3).
3. Ibid p.28
4. W. Foy (ed.), *Man's Religious Quest; A Reader.* Croom Helm, London, 1978, p.402f
5. David Goldstein, op. cit. p.35
6. N. Smart, op. cit. p.362
7. Hans Kung, *Judaism.* SCM, London, 1992, p.595ff
8. David Goldstein, op. cit. p.69. The Orthodox hold that there will be a physical resurrection in the Messianic age but the Liberal and Reform Jews deny it, though they do accept the immortality of the human spirit. Belief in bodily resurrection differs from the Greek view of the immortality of the soul. The Jews are a practical people and enjoy the material world while recognising that its values can delude. Salvation relates to their nation rather than the individual. If God is obeyed it will lead to his kingdom on the earth rather than an undefined spiritual 'world to come'.
9. N. Smart, op. cit. p.366. God unfolds the divine Nothing, En Sof (the infinite). We cannot know this hidden aspect. Other aspects are unfolded in the divine becoming, and provide the unification of God. Adam's sin disrupted the unity in God so humanity affects him and mankind must strive to promote the wholeness of God. The tenth sefirah is also called the Shekhinah, that is, glory or presence

of God which is used in the Jewish Targums to signify God himself. The similarity with emanation from Brahman is clear and as in Sankara worship and prayer is on the lower level. While these views have been influential, they have not been central to the Jewish faith and its modern successors, the Hasidim, have become more orthodox.

10. R. C. Zaehner (ed.), *Concise Encycolopedia of Living Faiths.* Hutchinson, London, 1979, second impression, p.16f. Maimonides produced 13 articles of the Faith: existence of God, unity, incorporeality and eternity of God, obligation to worship God alone, prophecy, the superiority of the prophecy of Moses, the Torah is God's revelation to Moses and is immutable, God's omniscience, reward and punishment, the coming of the Messiah, and the resurrection of the dead. p.31ff

11. See K. Ward, *Images of Eternity.* Darton, Longman and Todd, London, 1987, p.97. H. Kung, op. cit. p.39. R. C. Zaehner, op. cit. p.410

12. Ibid pp.91f, 123

13. A. Konig, *Here I am.* UNISA, Pretoria, 1978, p.78ff. Konig points out that these two limitations are of great importance. First of all because they are so peculiar to our human life that they would without any doubt have been applied to God if God were a product (projection) of Israel's thinking and longing (Feuerbach). Secondly, because in the religions around Israel and indeed in almost all heathen religions the regular death (or sleep) and resurrection of the gods, as well as their sexual lives, featured prominently.

14. D. Bohm, *Wholeness and the Implicate Order.* Routledge & Kegan Paul, London, 1980, p.177

15. See A. Sharma, *A Hindu Perspective on the Philosophy of Religion.* Macmillan, London, p.67f

16. But the kind of picture of Jesus that emerges depends on the approach of critical research: historical, literary, source criticism. These approaches are well known but we mention R. A. Burridge who has recently argued, unlike the Form critics, that the genre of the Gospels imply a biography and he compares them with such work in the ancient world. *Four Gospels, One Jesus? A Symbolic Reading* (SPCK, 1994). The approach gives the theologian more hope of basing his work on what the Gospels tell us about Jesus.

17. O. Hanfling, *The Quest for Meaning.* Blackwell, Oxford, 1987

18. J. Baker, *The Foolishness of God.* Darton, Longman & Todd, London, p.252f

19. G. Bostock, 'Do we need an Empty Tomb?', *The Expository Times*, Vol. 105, No. 7, p.201ff, April 1994. He suggests that it was the Jewish religious leaders who removed the body since they were afraid that his grave would become a martyr's shrine. They then alleged that it was the disciples who had removed it in order to explain why the tomb was empty. But we may ask: could they not have produced the body? To reply that it was too embarrassing for them to admit that they had stolen it is not convincing. Or to say that the body would have decomposed by Pentecost and could not have been produced appears

equally invalid, for rumours must have circulated before then and needed quashing.

20. Quoted by K. Armstrong, *Muhammad.* Gollancz, London, 1991, p.257

21. John Wolffe, 'Fragmented universality: Islam & Muslims' in *The Growth of Religious Diversity: Britain from 1945* ed. by G. Parsons vol. 1. *Traditions*, OUP & Routledge, London, 1993, p.166. Rushdie insinuated that Muhammad by his manner of life was not a reliable vehicle for divine revelation and accused him of making use of the so-called satanic verses (recorded by early Islamic authorities but rejected by others), which sanctioned the veneration of other gods (Meccan). He did it in order to make himself popular with the people. Rushdie said that these were not his views but that of the character in the novel but he was unable to refute convincingly the Muslim criticism.

22. N. J. Dawood, *Koran.* Penguin Books, Middlesex, 1956, pp.154, 331, 370. The Muslim believes that the Qur'an is the word of Allah with its beautiful language and inspiration. Translation into another language loses much of the effect but the Arabic recitation induces the feeling that Allah is present. While some Muslims would contend for the use of human faculties in the receiving of the revelation, the general view is verbal inspiration given to the prophet in a state of receptivity (wahy). There was a gradual development in the collecting of what had been revealed to Muhammad with final completion occurring in the reign of the second and third Caliphs (prior to 656 AD).

23. K. Gragg, *Jesus and the Muslim.* Allen & Unwin, London, 1985, p.293. *Islam and the Muslim.* Open University, Milton Keynes, 1987, pp.14, 67.23. There are five pillars of Islam: witness (shahadah) to Allah, and recognition of his prophet and prayer (salat), zakat (alms giving), fasting (siyam) and pilgrimage (hajj). The Muslim revers Mecca and the sacred shrine of the Ka'bah. It is set in the courtyard of the Mosque and has in the east corner the Black Stone which tradition holds was built by Abraham and Ishmael. Since the Qur'an is Allah speaking it is the presence of God in the Muslim community.

24. M. A. Salami, *Muhammad, Man and Prophet.* Element, Dorset, 1995, p.74. He forgave the elders of the city of Mecca who had plotted and fought against him. When, after their defeat, they stood before him with bowed heads he said: 'May Allah pardon you. Go in peace. I say to you as Joseph said to his brothers. There shall be no responsibility on you today. You are free.' Muhammad is a model of wisdom and trustworthiness, exemplified in his action when it came to fixing the sacred Black Stone in the wall. All the Meccan leaders wanted the honour of fixing the stone and a quarrel arose among them. But Muhammad was selected as a judge and he spread a white sheet on the ground and placed the Black Stone in the middle. The tribal leaders were asked to hold the sheet and then carry it to its site. Muhammad fixed the stone in its place. Such a cooperative effort satisfied all the parties and earned Muhammad the title of Al Amin, 'the trustworthy'.

CHAPTER 9

1. I. Barbour, *Myths, Models, Paradigms*. SCM, London, 1974, p.30f
2. I. Barbour, *Issues in Science and Religion*. SCM, London, 1966, p.154
3. A. Peacocke, *Science and the Christian Experiment*. Oxford University Press, 1971, p.12f
4. W. Van Huyssteen, 'Theology and Science, the quest of a new apologetics', *Princeton Seminary Bulletin* No.2, 1993, p.113f
5. I. Barbour, op. cit. p.443f
6. William Barclay, *Gospel of Luke*. St Andrews Press, Edinburgh, 1953, p.38
7. Karl Barth, *Church Dogmatics*. T & T Clark, Edinburgh, 1957, p.133f
8. J. Hick, *An Interpretation of Religion*. Macmillan, 1989, p.176ff
9. W. Foy, *Man's Religious Quest*. Open University, 1978, p.174
10. The proposal to reduce the significance of Christ stands in marked contrast to what Gandhi said about him: 'The lives of all have, in some greater degree, been changed by his presence, his actions and the word spoken by his divine voice ... And because the life of Jesus has had the significance and the transcendence to which I have alluded, I believe that he belongs not solely to Christianity but to the entire world, to all races and people; it matters little under what flag, name or doctrine they may work, profess a faith or worship a God inherited from their ancestors.' Of course Ghandi with his Indian background did not restrict incarnation to Christ.
11. G. Knight, *A Biblical Approach to the Doctrine of the Trinity*. SJT Occasional Papers No.1, 1957, Oliver & Boyd, Edinburgh, p.63
12. C. Gunton, *Unity and Diversity, God World and Society*. SST 1993 Conference, Cardiff, p.6
13. Ibid p.7
14. C. Schwobel, *Christology and Trinitarian Thought*. SST 1993 Conference, p.10
15. W. Johnson 'The Doctrine of the Triune God Today', *Insights*, Austin Presbyterian Theological Seminary, Texas, Fall 1995, p.6ff. Hegel saw God as realising himself in the world as Absolute Spirit. In the movement God is dynamically personal, relational and agential, becoming in revelation other than himself for self-differentiation needs to actualise itself in the world. There was no time when God was separate from the world so he rejects a creator, and it has no relation to the person and work of Christ. Karl Barth stresses that God is in revelation as he is in his eternal being. There is a single I in a threefold structure or three modes of Being. This contrasts with what others hold: three divine and co-equal persons which emphasise the personal nature of God. Moltmann, Jenson and Rahner point to three distinct centres of consciousness as against Augustine, but persons are not understood in the modern sense. God is not a finite individual or group of individuals but 'the eternal and non-contingent source of personhood.'
16. A. Plantinga, *Concept of God*. Ed. T. V. Morris, OUP, p.199. Vardy comments on an analogy of Peter Geach who likens God to a Grand Chess master and we as chess players are free to make any move we want. But it is obvious considering the nature of our opponent that we are going

to lose. God cannot be hindered ultimately by the moves that are made. But it seems to interfere with our freedom hence it is better to say that while a Grand Master will beat us for he knows all the possibilities open to us yet he does not know the specific choices that we will make. What he knows is that his ultimate plan will not be frustrated. Knowing what a person will do does not mean determining him or her in a causal way. P. Vardy, *The Puzzle of God*. HarperCollins, London, 1990, 17f.

17. J. Polkinghorne, *Reason and Reality*. SPCK, London, 1991, p.82

CHAPTER 10

1. W. L. Craig, *The Cosmological Argument from Plato to Leibniz*. Macmillan, 1980, p.152. Cf. Ninian Smart, *The Religious Experience of Mankind*. Collins, London, 1969, p.385
2. David Scully, *God and Reason*. CEM, London, p.51
3. John Hick (ed.), *The Existence of God*. Macmillan, New York, 1973, p.9
4. Paul Davies, *God and the New Physics*. Dent, London, 1983, p.168
5. Ian Barbour, *Religion in an Age of Science*. SCM, London, 1990, p.25
6. Ibid pp.7, 25. Hume pointed to the suffering of the world which implied that either God was not good or unable to do anything about it. The world is imperfect so it must be the work of an inferior deity. Religions however explain the imperfection of the world by the entry of sin and liberation from it will mean a new age of perfect design in which 'the wolf will live with the lamb, the leopard will lie down with the goat' (Is.11.6–9). The liberation of mankind will be accompanied by an answer to the groanings of nature (Rom.8.22–3).
7. A. R. Peacocke, *Theology for a Scientific Age*. SCM, London, 1993, p.119ff
8. John M. Templeton, *Evidence of Purpose*. Continuum, New York, 1994, p.114
9. K. Ward, *God, Chance & Necessity*. One World, Oxford, 1996, p.172
10. Proslogion 2. p.117 cited by Ward. Ibid p.60
11. John Hick, op. cit. p.63
12. N. Smart, *Philosophers and Religious Truth*. SCM, London, 1969, p.25f. We might make a point concerning the naturalist interpretation of miracles, that is, that they still show the amazing insight of Jesus. In the story of the widow's son who seemed to be dead Jesus approached the bier and told the young man to rise. It is possible, since the climate demanded quick burial, that he was in a cataleptic trance and Jesus saved him from being buried alive. There is evidence from the tombs in Palestine that many were buried alive. In the story about the daughter of Jairus, the president of the synagogue, Jesus said: 'She is not dead but sleeping', which provoked the laughter of the crowd. In both cases there is a miracle of diagnosis which shocks and surprises the people (Mk.5.22–43; Lk.7.11–16).
13. Michael Fuller, *Atoms and Icons*. Mowbray, London, 1995, p.32

14. Russell Stannard, *Doing Away with God*. HarperCollins, London, 1993, p.122. We would stress that miracles must be scrutinised for there is much credulity. Village Hinduism has a belief in miracles and there have been reports of statues drinking milk in temples similar to stories of the statue of the crying Virgin Mary. The Hindu philosophers admit that their faith includes miracles but point out that the stress is on moral truths. However, the abuse of miracle does not mean that it is false just as the abuse of belief in God does not prove his non-existence.

15. Fuller, op. cit. p.25. Pascal's account of an overwhelming experience of God was discovered after his death written on a piece of paper that was sewn into his clothing. It occurred on the 23rd November 1652, from half past ten until half past midnight, and reveals a feeling of nothingness before the awe-inspiring object, a sense of guilt for past misconduct, and a vow to be devoted to God in the future.

16. T. S. Kuhn, *The Structures of Scientific Revolutions*. University of Chicago Press, 1970, p.52ff

17. Arvind Sharma, *A Hindu Perspective on the Philosophy of Religion*. Macmillan, London, 1990, p.108

18. N. Smart, *The Religious Experience of Mankind*. William Collins, Glasgow, 1969, p.366f

19. Freud pointed to the root of religion in childhood experiences and the relation to sex. The son resents the father's possession of the mother and unconsciously desires to kill him and the daughter wants to remove the mother in order to have the father. Such feelings are repressed and overcompensation occurs so that the son idolises the father and makes him into a God figure. Religion is the projection of the father image and is an illusion. Freud's view reflects the experience of neurotic patients, the passion for his youthful mother, and jealousy of his father. The theory does not apply to the normal family life and many examples of religious experience do not have any connection with sex. Jung thought that Freud's ideas about the personal unconscious did not reveal all the factors involved or explain the bizarre symbols that appeared in dreams. He put forward a Collective Unconscious to account for the similarity of myths, and dream symbols among different classes of people. Neuroses are the outcome of spiritual problems that religion can resolve. When asked if he believed in God he replied: 'I do not have to believe I know!' But Jung was not orthodox in his model of God since he saw the shadow of evil in the deity.

20. S. Brown, 'Secular alternatives to Religion', *Man's Religious Quest* Units 22–23, OU Press, Milton Keynes, 1981, p.73

21. Peter Berger, *A Rumour of Angels*. Penguin, London, 1970, ch.3

22. A. J. Ayer said in his debate with Copleston that he did not want to re-strict experience to sense-experience for people might have religious experiences which he did not have. *The Meaning of Life and Other Essays*. Weidenfeld & Nicolson, London, 1990, p.37

23. I have cited this experience in *Can we ever Kill?* HarperCollins, London, 1991, p.189ff. Now published in an expanded second edition by Darton, Longman and Todd, London 2000.

24.   R. B. Braithwaite insisted that we look for the use of rather than the meaning of statements. Just as we cannot rule out electricity because its current is unobservable we cannot say that values do not exist because we cannot describe them as we do colours or the weights of objects. Science has many abstract concepts far removed from the kind of verification that A. J. Ayer demanded in his early work.
25.   D. Cupitt and D. Z. Phillips, *Is God Real?* ed. by J. Runzo, Macmillan, 1993, London, p.199f. To say that there is no objective reality corresponding to God avoids the problem of transcendence but in what way does it differ from humanism? There is a dispute between Phillips and Cupitt regarding language as how it might refer but both appear to make God into a personification of our ideals. It is possible to pay attention to language and yet hold with God as a transcendent Reality as seen with Barth. With him, theology is similar to what Wittgenstein calls logical 'grammar' for it seeks to elucidate the conceptual structure of biblical language and reveal the inner logical connections. The tests to be applied to any theological statement are internal: scripture, the community, and so on. But we cannot be content with a plurality of unrelated languages and Wittgenstein did speak of family resemblances as well as family differences.
26.   S. T. Davis, Ibid p.58. Phillips finds fault with those who mount arguments for the existence of God and maintain that such existence is probable (Swinburne) or cumulative (Mitchell) or more likely than contrary arguments (Plantinga). But arguments for his existence though limited do have value in suggesting an intention and purpose behind the universe. If our analogies reach some partial truth about him they are worthwhile and the Spirit promises to lead us into truth (Jo.16.12–15). However, we need to test the spirits (1 Jo.4.1–3).
27.   B. McGuinness, *Wittgenstein: A Life: Young Ludwig (1889–1924)*. Duckworth, London, 1988, pp.255, 273

CHAPTER 11

1.   J. Robinson, *Exploration into God*. SCM, London, 1967, p.130
2.   D. M. Baillie, *God was in Christ*. 1955
3.   K. Cragg, 'Islam and the Muslim', *The Religious Quest*. OU Press, Milton Keynes, 1987, p.64, units 14–154
4.   A. A. Jones, *Man's Religious Quest, A Glossary of Terms*. OU Press, Milton Keynes, 1978, p.40
5.   D. Hardy, 'Theology, Cosmology and Change', SST Oxford Conference 1994, p.19
6.   A. Peacocke, 'Science and the Theology of Creation', Society for the Study of Theology p.10
7.   Ibid p.9
8.   A. R. Peacocke, *Creation and World of Science*. OUP, 1979, p.142
9.   C. H. Dodd, *Romans: Moffat New Testament Commentary*. London, 1932, p.113. J. Burnaby, *The Belief of Christendom*. 1960, p.21

10. Paul Davies, *New Scientist*, 27 April 1996, No. 2027, p.32
11. Ibid p.35
12. G. Walker, *New Scientist*, Ibid p.35
13. K. Ward, *God, Chance and Necessity*. One World, Oxford, 1996, p.40.
    M. Fuller, *Atoms and Icons*. Mowbray, London, 1996, p.77
14. R. Stannard, *The Times*, 13 November 1989
15. P. Strawson, *Individuals*. Methuen, London, 1959, p.115
16. Lady Oppenheimer, 'Spirit and Body: the prophetic church and the problem of corporate agency'. SST Oxford Conference 1992, p.20
17. K. Ward, op. cit. p.55
18. J. Badcock, *Elements of Christian Symbolism*. Element, Dorset, 1990, p.34f
19. J. S. Stewart, *A Man in Christ*. Hodder & Stoughton, London, 1935, p.167
20. J. Polkinghorne, *Scientists as Theologians*. SPCK, London, 1996, p.55
21. M. Fuller, op. cit. p.36
22. J. Polkinghorne, op. cit. p.45ff
23. R. Stannard, *Science and the Renewal of Belief*. SCM, London, 1982, p.172
24. R. C. Zaehner (ed.), *Concise Encyclopedia of Living Faiths*. Hutchinson, London, 1979, p.226
25. R. Harries, *The Real God*. SCM, London, 1994, p.64
26. William Rankin, *Newton*. Icon, Cambridge, 1993, p.106
27. J. G. Brunton, *The Story of Western Science*. CUP, Cambridge, 1966
28. R. Harries, op. cit. p.8
29. K. Ward, op. cit. p.201ff

Index

231